Contributions To Phenomenology

In Cooperation with The Center
for Advanced Research in Phenomenology

Volume 97

Scope

The purpose of the series is to serve as a vehicle for the pursuit of phenomenological research across a broad spectrum, including cross-over developments with other fields of inquiry such as the social sciences and cognitive science. Since its establishment in 1987, *Contributions to Phenomenology* has published more than 80 titles on diverse themes of phenomenological philosophy. In addition to welcoming monographs and collections of papers in established areas of scholarship, the series encourages original work in phenomenology. The breadth and depth of the Series reflects the rich and varied significance of phenomenological thinking for seminal questions of human inquiry as well as the increasingly international reach of phenomenological research.

The series is published in cooperation with The Center for Advanced Research in Phenomenology.

More information about this series at http://www.springer.com/series/5811

Natalie Depraz • Anthony J. Steinbock
Editors

Surprise: An Emotion?

Springer

Editors
Natalie Depraz
Department of Philosophy
University of Rouen
Mont Saint Aignan, France

Anthony J. Steinbock
Department of Philosophy
Southern Illinois University Carbondale
Carbondale, IL, USA

ISSN 0923-9545 ISSN 2215-1915 (electronic)
Contributions To Phenomenology
ISBN 978-3-319-98656-2 ISBN 978-3-319-98657-9 (eBook)
https://doi.org/10.1007/978-3-319-98657-9

Library of Congress Control Number: 2018960386

This Springer imprint is published by the registered company Springer Nature Switzerland AG
The registered company address is: Gewerbestrasse 11, 6330 Cham, Switzerland

Introduction

The contributions that are gathered in the present volume were first given as talks during a conference entitled "Surprise, an Emotion?" that took place at Southern Illinois University, Carbondale, on September 25–27, 2013.

It was co-organized as a collaborative colloquium by N. Depraz and A. Steinbock and jointly sponsored by the Phenomenology Research Center at SIUC, the SIUC College of Liberal Arts on the one hand, and the ANR (Agence Nationale de la Recherche) EMCO (Emotion-Cognition)-Emphiline Research Program (2012–2015) entitled "Surprise at the core of the Spontaneity of Emotions: an Extended Cognition," the Centre de la Recherche Scientifique (CNRS) and the Archives-Husserl de Paris at the Ecole Normale Supérieure (UMR Pays Germaniques) on the other. It was also generously sponsored by Springer Publishers to produce this edition. We want to give our heartfelt thanks to all of these institutions, without which such a conference and book edition would not have been possible.

Surprise is a theme that has not been dealt with directly and systematically in philosophy, in sciences, in linguistics or in spiritual traditions, even though many philosophers of different traditions, a number of experimental studies in the last decades, recent works in linguistics and ancestral wisdom testimonies mention it as a crucial experience of both rupture and openness in bodily and inner life. Thus, it is often described thanks to some closely related words, concepts and exclamations-interjections such as startle, astonishment, event, wonder, whoa! or ah!, but very rarely thematized at length and in its full meaning.

As a unique predecessor presenting a pioneering integrated psychological model of surprise taking into account its subjective component, let us just mention the Bielefeld School, which was particularly active during the 1990s in the twentieth century, namely, around the psychologist Rainer Reisenzein. Following up the 40th Conference of the *Deutsche Gesellschaft für Psychologie*, organized in 1996 in Munich, N. Frijda edited a volume entitled *Proceedings of the 9th Conference of the International Society for Research on Emotions* (Toronto, Canada, 1997). An initial inaugurative article was in this volume, one jointly published by R. Reisenzein,

W. U. Meyer and A. Schützwohl under the title, "Analyse von Reaktionen auf überraschende Ereignisse: Ein Paradigma für die Emotionsforschung."[1]

The ANR Research Project Emphiline "Surprise at the core of the spontaneity of emotions: an extended cognition" (2012–2015) conducted by Natalie Depraz takes up the issue of surprise anew and wishes to study it under the light of multidisciplinary approaches: philosophy, psychophysiology, psychiatry and linguistics. Some first collective volume editions begin to reveal the importance of the question, either at the interface of philosophy and linguistics,[2] at the crossroads of phenomenology and psycho-neuro-physiology[3] or again in philosophy-phenomenology alone.[4]

As a transversal issue, it is our insight that surprise may be best broached and examined through an interdisciplinary lens; this alone allows for a genuine enriched descriptive understanding of a complex and multidimensional intentionality: the active-passive, i.e. receptive phenomenon of both surprising and being surprised.

The present volume is an attempt to offer some initial perspectives on the theme of surprise crossing philosophical, phenomenological, scientific and linguistic contributions. It has been made possible thanks to the long-standing friendship and shared phenomenological interest of both editors, each of whom has developed previous research and decided to share them: Anthony Steinbock is the director of the Phenomenology Research Center at SIU and has dedicated many years to the study of emotions (pride, shame, guilt, repentance, hope, despair, trusting, loving, humility). He recently published *Moral Emotions: Reclaiming the Evidence of the Heart*.[5] Following his book, *Phenomenology and Mysticism: The Verticality of Religious Experience* (Indiana UP, 2007/2009), he draws on his theory of personhood and inter-personality and focuses on moral emotions, distinguishing in them three main modes of givenness: emotions of self-givenness, emotions of possibility and emotions of otherness, respectively. Natalie Depraz, as the director of the ANR Emco-Emphine, chose as its leading theme the phenomenon of surprise; she explores it in a multidisciplinary way, jointly with a research seminar at the Husserl-Archives (ENS/CNRS) about emotions generally (2012–2015) and more specifically about

[1] N. Frijda (ed.), *op. cit.,* R. Reisenzein, W.-U. Meyer, & A. Schützwohl, « Reactions to surprising events. A Paradigm for Emotion Research », in N. Frijda (ed.), 1997, *Proceedings of the 9th Conference of the International Society for Research on Emotions,* Toronto, Canada, 1997, pp. 292–296. Some other articles were published in the following years (see Depraz' article in the present volume), but it seems to be that broader manuscripts remained then unpublished.

[2] N. Depraz and Cl. Serban eds., *La surprise dans les langues*, 2015, Paris, Hermann, and N. Depraz and A. Celle, *Surprise at the intersection of phenomenology and linguistics*, Amsterdam, J. Benjamins Press, 2018.

[3] Th. Desmidt, M. Lemoine, C. Belzung & N. Depraz, « The temporal dynamic of emotional emergence », *Phenomenology and the Cognitive Sciences, Emotion Special Issue*, 2014, Springer, Heidelberg; N. Depraz & Th. Desmidt, « Cardio-phénoménologie », in: *La naturalisation de la phénoménologie 20 après*, J.-L. Petit ed., Cahiers philosophiques de Strasbourg n°38, 2015.

[4] "La surprise", *Alter* n°24, Revue de phénoménologie, Paris, 2016, N. Depraz and V. Houillon eds.

[5] A. J. Steinbock, *Moral Emotions : Reclaiming the Evidence of the Heart* (Evanston, IL: Northwestern University Press, 2014).

aesthetic and traumatic emotions (2015–2016) and in the framework of a translation group at the Husserl-Archives co-organized with Maria Gyemant dedicated to Husserl's *Studien der Struktur des Bewußtseins* and, more particularly, focusing on *Gefühl*, *Wunsch* and *Begierde*.

Is Surprise an Emotion? Such is the leading question of the present volume. Two main theoretical frameworks are available here to help us let the specificity of surprise emerge. In psychology, surprise is commonly considered as a primary emotion; in philosophy, surprise is related to passions as opposed to reason. Are such entries into surprise satisfying or sufficient? To what extent is surprise also a cognitive phenomenon and primitively embedded in language, be it expressive or inserted in the very structure of language? How is surprise opened to personhood, the interpersonal and moral emotions?[6]

Mont Saint Aignan, France — Natalie Depraz
Carbondale, IL, USA — Anthony J. Steinbock

[6] We would like to express our gratitude to Aaron Darrisaw for his unflagging and generous help in preparing this manuscript for publication.

Contents

Part I
Surprise and the Heart

Surprise as Emotion: Between Startle and Humility

Anthony J. Steinbock

Abstract I consider the experience of surprise within the context of my current work on the emotions. To do this, I examine surprise in terms of its belief structure, distinguishing it from a startle (1). I then suggest that surprise is a being caught off-guard that is related to being attentively turned toward something (2). As the latter, I qualify surprise as an emotion in its being thrown back on an experience in a way that is different from affectively turning toward something (3). This constitutes surprise as a disequilibrium in distinction to a diremptive experience like we find in the moral emotions of shame or guilt (4). Finally, I distinguish surprise from a gift, which is peculiar to the experience of humility. I then suggest that surprise is an emotion while being neither an affect, like a startle-reflex, nor a moral emotion, like shame, guilt, or humility.

Keywords Surprise · Startle · Humility · Gift · Phenomenology

Although the experience of surprise is prevalent in everyday experiences and seems to be self-evident, it is a distinctive experience that is anything but clear. Indeed, while many philosophers and scientists of different traditions do mention it as a crucial experience, surprise as a theme has not been dealt with systematically in philosophy (or in the sciences). In contemporary literature in which the gift becomes a topic of discussion, the gift is commonly and simply assumed (somehow) to be coupled with surprise. In this chapter, I am interested in how surprise can considered to be an emotion, its difference in relation to startle and humility. This, in turn, concerns how surprise is related to a gift, and the essential, crucial distinctions between them.

To do this, I want to ask: Does surprise have an affinity with perceptual and general epistemic functions and acts? What is its relation to the future? Does it have an epistemic import? Is surprise an affect? An emotion? What is surprise's relation to moral or interpersonal emotions? It would be too ambitious to respond in detail to all of these questions. But I do want to bring them into focus by determining surprise within the problem-field of feeling and then situate it in relation to the gift.

A. J. Steinbock (✉)
Department of Philosophy, Southern Illinois University Carbondale, Carbondale, IL, USA

N. Depraz, A. J. Steinbock (eds.), *Surprise: An Emotion?*, Contributions To Phenomenology 97, https://doi.org/10.1007/978-3-319-98657-9_1

More specifically still, I examine surprise in terms of its belief structure, clarifying it as a believing what I cannot believe, and ultimately distinguishing it from a startle (1). I then suggest that surprise is a being caught off-guard, which is related to being attentively turned toward something (2). As the latter, I qualify surprise as an emotion in its being thrown back on an experience (3). This constitutes surprise as a disequilibrium in distinction to a diremptive experience like we find in the moral emotions of shame or guilt (4). Finally—and contrary to a common interpretation—I distinguish a surprise, which presupposes an expectation, from a gift, which is peculiar to the experience of humility and which (while it has its own futural temporality) is that in which precisely nothing is expected. I then suggest that surprise is an emotion while being neither an affect, like a startle-reflex, nor a moral emotion, like shame, guilt, or humility.

1 The Belief Structure of Surprise

Surprise can be characterized by a peculiar relation to being. Allow me to describe this relation by examining its "belief structure," especially where the future is concerned. I do this because it is commonly held that surprise is simply a rupture of what is expected.

1.1 Expectation: Acceptance of What Is to Come

Within the phenomenological tradition, we discern temporal modes of time-consciousness relating to the present, the past, and the future. Where the future is concerned, we can observe a "protention," or an anonymous sketching out of the future that is based on a present occurrence and how that occurrence was retained as past. This takes place without any egoic activity or explicit attention to what is to come; it takes place through the "passive synthesis" of sense. An expectation is similar to a protention insofar as it is open to a futural occurrence arriving in the present, and it is also unfurled from the present and the past. Expectation is different from protention, however, insofar as expectation is an active comportment toward the future. In relation to this, we can see how an anticipation can be a more intensive attentiveness to the futural event.

For example, as I run to Times Square on New Year's Eve, my steps protend an even pavement; I place one foot in front of another without even thinking of it. All of this happens as I expect the taxi to come to a stop so that I can cross the street; and this takes place as I anticipate with bated breath the ball to drop in Times Square. These are all distinctive orientations toward the future, even though they all may be lived simultaneously. In order now to discern how surprise is dependent upon this futural orientation in the latter's various modes, let me describe the belief structure inherent in such a futural temporal mode of givenness.

For the sake of simplicity, let's stay with an expectation. Intrinsic to the act of expectation is the fact that the existence of something futural is posited. Expectation is carried out in the mode of belief as an unbroken, straightforward relation to something in the future. When I see the police car in my rearview mirror with its lights flashing (after I know that I have been going too fast), I expect the police car to pull up behind me and signal for me to pull over. When I expect this, I implicitly posit the existence of the officer, the police car, the lights, the forthcoming ticket, etc. That is, the being of the officer, and so forth, is accepted in terms of the sense or meaning it has as going to come to pass; I live in the mode of natural, straightforward acceptance. This is another way of saying that when I expect something, I expect it as actually going to happen, not as something possible, or as possibly going to happen.

When I see the police car I "posit" or accept it as actually behind me and as going to pull me over; when it speeds past me, going after the car in front of me, I accept with relief! its actual passing, its "going to pull over that car." Further, expectation in all of its forms is not a rupture of belief; it is another kind of belief as a mode of time-consciousness, a straightforward one oriented in the direction of future actuality. In expectation, I count on the futural event as it is foreshadowed in the present. Thus, expectation is a temporal belief-act that is oriented toward the future as a mode of time-consciousness; it arises as motivated on the basis of, is demanded by what has occurred in the present and is immediately retained as past.

Now, there are ways in which this straightforward futural acceptance can be modified or modalized. For example, something can be given as *possibly* going to occur, as *likely* to happen, as *probably* going to arrive, and so forth. For instance, I believe that the experiment to confirm the existence of the Higgs boson will probably work. Or, if there are too many counter-indicators—let's say that the equipment malfunctioned—I am *doubtful* it will work this time. The point here is that likelihood, possibility, probability, even doubt, are all kinds of belief postures or modes of belief. Aristotle seems to place the phenomenon of surprise here. In the *Poetics*, Aristotle connects *ekplektikon/ekplexeos*—which can be translated as "surprise" or being *struck* in "awe" or "astonishment"—with subjective discoveries through probable incidents.[1] *Thaumaston*, from *thaumazein* or wonder, on the other hand, is more open for the improbable; in fact, it is produced through improbable (or unexpected) incidents in relation to one another, which yields a great epistemic effect.[2] Wonder [*thaumaston*] for Aristotle, then, is a different phenomenon and cognitively "higher" than surprise [*ekplektikon/ekplexeos*]—both related to the probable and improbable.

[1]Aristotle, *Poetics*, ed., and trans., Stephen Halliwell (Cambridge: Harvard University Press, 1995), 1454a 4; 1455a 17. Regarding wonder [*to thaumazein*] as the beginning of philosophy, see Aristotle, *Metaphysics*, Books I-IX, trans., Hugh Tredennick (Cambridge: Harvard University Press, 1933), 982b 12–17.

[2]Aristotle, *Poetics*, 1452a 4–6; 1460a 12–14. See 1460a 26–27: "Things probable though impossible should be preferred to the possible but implausible." Impossibilities are justifiable if they make that portion of the work more astounding (1460b 23–29).

Initially we might want to say that surprise is the experience of the unexpected, and in this way it may well sever its relation to belief. Adam Smith, for example, distinguishes between three "sentiments" that can initiate philosophical inquiry: wonder, surprise, and admiration. Wonder is excited by the new and novel; admiration is provoked by the beautiful; surprise is motivated by the unexpected, which for him is tied to the sudden, but not the rare.[3] Where Edmund Husserl's phenomenology is concerned, Smith's surprise would be a "disappointment," whereas Smith's wonder would be a "discordance."[4]

1.2 I Believe What I Can't Believe

Let's return to the experience of surprise in relation to an expectation. How do we characterize its peculiar belief-structure? When we are surprised, it often feels like an "I can't believe it," an "I can't believe what has just happened," or an "I can't believe what is happening." This experience entails a being caught off guard. This is why we can even be surprised in relation to ourselves (i.e., I can also surprise myself): "I can't believe what I just said" (say, I lost my temper, but I never lose my temper!). Or, "I can't believe what I just said; I know I said it" (I accept it), "but it goes against what I expect of myself."

The expressions given above that we find in our everyday experience are important clues to the experience of surprise. But if we were to remain simply with this aspect of the experience it would conceal the deeper process of the constitution of sense in surprise. By "belief," I understand a basic "doxic" attitude, posture, or disposition that accepts the being of what arrives, has arrived, and what is to come in a straightforward manner. In this respect, "belief" does not have to be an active reflective commitment to or positing of being, but it can be a "pre-supposing" or "passive-positing" of being in a kind of pre-predicative "taking in" what takes place or what is.

With this understanding we can more accurately portray what is happening in surprise. Surprise can be characterized as a movement of an "I am now believing what I could not believe at first," or again, "I am somehow accepting what I can't (in other circumstances) accept," or "I am living what I did not expect." In short, we have an "I believe what I can't believe," "I accept what I can't accept." For example, I never expected a birthday party, but here it is! Or more tragically, I can't believe she took her own life; she seemed so happy and successful, but she did commit suicide! In surprise, it is as if what happens comes out of nowhere, precisely because

[3] Adam Smith, *The Early Writings of Adam Smith*, ed., J. Ralph Lindgren (New York: Augustus M. Kelly, 1967), esp., 30–31, 33, 39. Wonder is an elaboration of surprise, concerning the singularity of the succession.

[4] See Edmund Husserl, *Analyses Concerning Passive and Active Synthesis: Lectures on Transcendental Logic*, trans., Anthony J. Steinbock (Dordrecht: Kluwer, 2001), esp. Part 2, Division 1.

it is otherwise than the expected flow or unfolding of what is to come. But in order for the event to be experienced *as* surprise, *I must still accept it*. Thus, in surprise there is an overall *reconstitution* or *reconfiguration* of sense where the event in question is concerned.

Accordingly, surprise, even on this descriptive level, is more complex and more "pre-reflective" in its doxic posture than, say, Donald Davidson or Daniel Dennett portray it.[5] Davidson is too judicative, holding that surprise is the realization that the previous belief was false, in the sense that there is an objective reality independent of previously held beliefs.[6] On the other hand, Dennett is partially correct when he writes that "Surprise is only possible when it upsets belief," but he does not specify how surprise is also the acceptance of this so-called "upset."[7]

If the "I can't believe" were entirely decisive, and not encompassed with an "I now believe what I didn't believe would happen," then we would have something like a *shock* instead of surprise. Here, the event would not be reconstituted and reintegrated in its belief-attitude. In this case, we would experience an "I cannot accept what I cannot accept." If there were not an acceptance of what I cannot otherwise accept—if the rupture of the straightforward relation to such a peculiar event were not accepted in some way—I would not live this experience as a surprise.[8]

Epistemically, we might want to classify surprise under the category of a disappointed perception, a disappointment that arises through a short-term or long-term rupture or discordance that is gradually reconstituted in sense. But due in part to its severity, intensity, and because in part it issues in an immediate reconstitution of sense, surprise is distinctive from a disappointment. Put in more Husserlian colorful terms, we could say that surprise is the experience of the "shattering" of the noema, the sense-content of my ongoing intentional acts. Such a shattering of the noema in surprise is exemplified in the film, *The Crying Game*: When Fergus (Stephen Rea) discovers that his new lover, Dil (Jaye Davidson) is a man, not a woman! What is demanded is a radical reconfiguration of sense, a new one supplanting the old (Husserl also writes of being "thrown from the saddle"). As is the case with any like reconfiguration of sense, the presence of the previous sense is retained, not erased, but as retroactively crossed-out in its very reconfiguration as it is held onto in the

[5] I would like to thank Natalie Depraz for bringing these works to my attention.

[6] Donald Davidson, *Rational animals. Dialectica* 36: 318–27 (1982). And Donald Davidson, *Problems of Rationality* (Oxford: Oxford University Press, 2004).

[7] Daniel Dennett, "Surprise, Surprise," Commentary on O'Regan and Noe, in *Behavioral and Brain Sciences* 24 (5), (2001), 982.

[8] I do not consider here the protraction of a shock or the relation of shock to trauma, how it can be repressed, etc. Nor am I considering the experience of grief as a possible subsequent response to surprise or shock.

Hegel never uses the term "Überraschung" in his *Phänomenologie*. However, for experiencing consciousness (not absolute Subject) every new structurally distinct encounter would throw this consciousness back on its experience such that it provides a motive for a step back or rather, discovery, and re-conceptualization. Perhaps "surprise" is a suitable term for the encounter of what seems incidental and alien to its own making (positing), rupturing its expectation, and being thrown back on experience, and then testing, experimenting, and eventually reconceptualizing it.

past in the retention or primary passive condition of remembering.[9] For this reason, in surprise we find constituted an acceptance of what occurs against all expectation. This is one reason why surprise is something other than a mere disappointed or discordant perception, where certain aspects are modified only to yield a coherent sense of the whole.[10]

Thus, we can discern two moments of the surprise-experience in relation to its belief structure. On the one hand, there is a "being caught off guard," a radical "otherwise" in relation to expectation and the unfolding of sense: an otherwise expressed in the "I can't believe," which is the noetic expression co-relative to the "shattering" or "explosion" of the noema or sense-content. On the other hand, there is precisely a belief, an acceptance of the very non-acceptance. In terms of the belief-structure, this rupture is more than a disappointment; nevertheless, this rupture is not decisive, since it is encompassed by a belief in what I otherwise could not believe at first.

1.3 Startle and Surprise

I would like to distinguish further between a startle and a surprise. In a still different example, we can imagine being in deep concentration while reading a book, then, suddenly someone who just came in let the screen door slam! I jump; I am startled. The startle is certainly a response to an unexpected givenness, a rupture, and it is sudden; but the startle in this instance takes place affectively *without any reconstitution of sense*. This has to do in part with the instantaneity of the startle. That is, if surprise is a believing what I cannot believe such that it entails a reconfiguration of sense, and shock is an I can't believe what I can't believe as a resistance to the reconstitution of sense for my meaningful world, then startle is neither of these. Accordingly, startle can be viewed in two registers, a static and a genetic one. A startle is characterized temporally with a suddenness, "now." We have two possibilities as to its constitution. On the one hand, it is neither reconstituted nor integrated in a flow of experiences—even if it presupposes a futural protention, which is why we can be startled in the first place! But, on the other, if the startle is viewed over time in the flow of experiences, it can be said to be "integrated" in the flow, but now as a "mere" disruption or rupture (or discordance) in the otherwise concordant or harmonious flow of meaning. But this does *not* mean that its presence demands a reconfiguration of the sense in order to be a startle.

Here, startle takes place under the threshold of the surprise and the shock. It is not a matter of not believing what I can't believe (shock), or believing what I can't believe (surprise). A startle takes place on a purely passive level of experience, and this is why it is appropriate to speak in terms of a startle-*reflex*.

---

[9] See Husserl, *Analyses*, esp. Part 2, §7.

[10] See Husserl, *Analyses*, esp. Part 2, Division 1. There is no feeling necessarily in the latter, no existential import.

The question of suddenness has been a tricky issue in traditional descriptions of the surprise phenomenon. For example, in *The Passions of the Soul,* Descartes considers wonder [*admiration*] to be the first among the six passions. When we judge something to be new or very different from what we formerly knew or from what we supposed that it ought to be, it surprises us, and this causes us to wonder.[11] Hence, Descartes can write that "wonder is a sudden surprise of the soul."[12] It is wonder that makes the soul consider the objects with attention—objects that seem to it rare or extraordinary. Notice that it is surprise that is even more "primary" than wonder and is itself founding for wonder. Further, Descartes' statement can be read in such a way that surprise is itself not essentially connected to the sudden. For Descartes, surprise arises from a judgment of something being new or quite different. So, while surprise for him is connected to novelty and founded in a judgment, it is ambiguously tied to suddenness. When surprise is sudden, it issues in wonder.[13]

For me this is a clue. While suddenness as a temporal experience can accompany surprise, suddenness need not accompany the rupture of experience in order to have an experience of surprise.[14] From a phenomenological perspective, it is startle that is essentially tied to suddenness, not surprise. Let's take another example, this time of the experience of a jack-in-the-box. I want to maintain that we are startled, not surprised, when we—especially when we were children—turn the crank of a jack-in-the-box: We know that "jack" is going to pop out, we turn and turn the crank, we wait and wait (we expect, without knowing precisely when), and then "pop!" out springs "jack." In fact, if "jack" does not pop out (something goes wrong), we will become disappointed (and maybe surprised, but not startled). This is similar to buying a ticket to go on a haunted house ride: We know we will be scared (or startled!), but we are not surprised by what happens. In fact, we expect it. Again, we might instead be surprised (and certainly disappointed) if the so-billed haunted house were just an ordinary open house tour listed by a real estate agent as a ploy to get us to see the house. Similarly, when the police car in back of me turns on her lights, I fully expect her to pull me over, turn on her lights, etc., but when she turns on her siren, I am startled.

---

[11] René Descartes, *The Passions of the Soul*, trans., Stephen Voss (Hackett Publishing Company, Inc., 1989), Art. 53. It is the first of six primitive passions: love, hate, desire, joy, sadness. See also, Sara Heinämaa, "Love and Admiration (Wonder): Fundaments of the Self-Other Relations," in *Emotional Experiences: Ethical and Social Significance*, eds., John J. Drummond and Sonja Rinofner-Kreidl (London: Rowman & Littlefield Int., 2018), 155–74.

[12] See Descartes, *The Passions of the Soul*, Art. 70.

[13] In order of foundation we would find: judgment, surprise, wonder, attention. Natalie Depraz concludes that wonder is *the emotional immediate after-effect* of surprise while attention is its cognitive after-effect. See Natalie Depraz, "Surprise, Valence, Emotion: The Multivectorial Integrative Cardio-Phenomenology of Surprise," this volume, chapter 2.

[14] It is also interesting to observe that a startle has an immediate bodily resonance, usually expressed in a kind of "jump," a violent twitch, or sudden reflex movement. While surprise is often expressed in raised eyebrows, a widening of eyes, or a stepping back, etc., I can still experience surprise without any such facial or bodily gestures.

Temporally speaking, therefore, we can say that while startle demands the temporal feature of suddenness, surprise does not. It is the suddenness of the jack-in-the-box that contributes to the startle, not the unexpected per se. By contrast, I can still be surprised without suddenness (contra Smith and possibly along with Descartes) or without the rareness or extraordinariness of the event (contra Descartes, but along with Smith). For example, it surprises me, and *continues* to surprise me that my friend acted in this way and that he continues to act the way he does, though there is nothing sudden about this. If fact, it may not be entirely unexpected. Discrepancy, contrast with, and interruption of the expected are not necessarily sudden. If my friend's actions no longer surprise me it is because they have been integrated into a familiar pattern of action.

In short, a surprise has to be understood as more than an experience of an unexpected givenness, though it entails that, and it must be more than a startle. Whereas a startle does not necessarily entail a reconstitution of sense, but can issue in an integration of sense; whereas it is sudden, and is a reflex (and can still remain passive in this respect), a surprise entails a reconstitution of sense, need not be sudden (it can have duration), and is minimally the initiation into egoic activity of some kind. In view of its epistemic import, surprise has been connected to discovery, wonder, and philosophical inquiry. But I think that there is more here. It also has *existential* import insofar as it allows us to go on when confronted with the blow of the unexpected. Rather than being stymied with the shock of the "I can't believe what I can't believe," surprise is the accommodation of us to the situation by being the acceptance of what I cannot accept. But this is not merely an epistemic posture; it does it through the sphere of *feeling*, which has its own style of cognition and evidence.

2 Being Caught Off Guard and Attentive Turning Toward

In the previous section, I considered surprise as a believing what I cannot believe, having placed it in the context of an acceptance of what is to come, and I distinguished surprise from a shock and a startle. Let me now examine what takes place in being caught off guard, and in accepting what I can't accept in the experience of surprise. Determining this will have important implications for whether surprise is a purely passive experience (passive in Husserl's sense), or whether it takes place in transition to or within an active, egoic sphere.

We can recognize that just because a passive protention is interrupted, this interruption does not constitute a surprise. I noted earlier that a protention is a passive, pre-egoic sketching out of the future that is based on a present occurrence and how that occurrence was retained as past. For example, as I take notes while reading a book, my bodily comportment is directed implicitly to what follows—from sitting on the chair, to the movement of my hands and eyes as I continue to read, (all of which may be fulfilled or disappointed by the oncoming events). Now, I go to take a sip of tea, reaching for my cup while still reading my book. The protentional threads of my hand guide me to where I last placed the cup. But I do not find it there.

All this can go on implicitly while still concentrating on reading a passage. Though the protention is functioning through and through without any explicit judgment or inference of this process, there is not necessarily the emergence of surprise if I do not find my cup. For instance, I may get frustrated and turn my attention to the cup and look around; at this point surprise has not necessarily emerged, either rupturing, calling into question, supplanting, or supplementing the temporal, kinesthetic process of the protention. Simply because things do not go according to how we expect them to run their course does not necessarily mean that it is surprising (or startling, for that matter.)

Furthermore, I can have a fully "passive experience" that is affectively significant without any explicit turning toward anything attentively. For example, I am writing or concentrating while reading a book, and there is a light buzzing sound. I am feeling more and more tense, but I do not become aware of this until after it hits a threshold; I finally notice that I am tense, bothered, or that the noise was getting under my skin, and that it was the annoying air conditioner. Looking back on the situation, I realize what was happening passively, and I note *now* that then I was "affectively" sensing the noise, but it never registered as such. Or, we can take a different example. Let's say that I am driving, and I am used to the repetitive thumping of my tires on the highway. These are affectively working on me, as it were; it becomes a protention of an irregularity that as a whole becomes so regular that its rhythm lulls me into expecting the irregular as regular, and I only notice it when it is no longer present.[15] When I do turn toward it attentively, I do so merely out of curiosity, but I am neither startled nor surprised.

Let me now address this last point, namely, the possibility of an attentive turning toward in the transition from passive synthesis of sense to a more "active" or "egoic" constitution of it. There are many kinds of such attentive turning-toward, but only one that pertains to surprise. Clearly, a startle *can* serve as a transition to an active turning toward. A loud bang, for example, can be so affectively prominent, that I immediately turn toward the door and notice it as such. Here we might have a movement from pre-constituted sense to actively constituted meaning of the door, where the door slammed shut, and the person walked through it.

But a rudimentary transitional experience can also found in recollection. For example, I am listening to my friend speak about a special plant, and then the mention of "fern" recalls to me, affectively, my time in the old growth forest, and I am called back actively in remembering to the ferns in the coastal forest. Accordingly, an affective turning toward something coming into relief affectively can instigate a new experience, and it can provoke the emergence of a new object.

What is distinctive about surprise in this regard is that it may not only serve as a transition from what is passively given to what is actively experienced; it not only instigates a new experience in the reconstitution of sense; but in being thrown off guard, *it throws me back on the experience*. I can examine it further, I can become

[15] See Husserl, *Analyses*, esp. §§37–39.

curious, but I do not have to do the latter in order for it to be a surprise. It is sufficient that in being turned toward the event, *I not only turn toward it*, but I am thrown back on that relevant segment of experience. This is an active turning toward and back on an experience through which surprise instigates egoic activity—even though the turning toward and being thrown back onto experience is not a voluntary effort on the part of the ego. In fact, being in control of the experience in this way would mitigate the very experience of surprise; indeed, the process of crossing-out and the reconfiguration of sense that does takes place, takes place passively, without my actively doing anything.

A basic point I would like to suggest here is that if surprise occurs, it can never occur as such on merely a "passive" level or as a mere rupture in pre-reflective passive associative concordance. It will either occur as a transition from a passive level to an active awakening, or function within the active sphere, issuing in an attentive turning toward. This also points to the passive level of experience on which the startle occurs, and the more active meaning-context in which surprise occurs. Thus, I distinguish between the being caught off guard peculiar to the surprise experience, and the startle, which is fully pre-egoic as an experience.

3 Surprise as an Emotion

Husserl has shown that everything coming into relief as a unity of sense does so as an affective relief and as exercising an affective allure on the perceiver.[16] It is not necessary to repeat these investigations here; it is sufficient to note that surprise shares with aesthetic and kinaesthetic experiences an *affective tonality*. But what kind of affective tonality does surprise have? Is it merely an affect? Does it exhibit a distinctive valence? Can it be considered to be an emotion, and if so, in what sense? Can it be qualified as what I have described elsewhere as a moral emotion?

In Kant's *Anthropology from a Pragmatic Point of View*, we have a very indirect invocation of surprise. That is, surprise is used first as a synonym for the disruption of expectation through contrast in order to clarify wit, but later it is pressed into service to define an affect. Because an affect is rash, as in the German term for surprise, *Überraschung*, it is distinct from a passion [*Leidenschaft*].[17] "Affect is surprise through sensation," whereby the presence of mind or self-control is suspended. Thus, surprise gives to a sensation "feeling" and "suddenness" to constitute the phenomenon of affect. Kant is therefore more ambiguous regarding the phenomenon

[16] See Husserl, *Analyses*, Part 2, Div. 3.

[17] Immanuel Kant, *Anthropologie in pragmatischer Hinsicht*, ed., Karl Vorländer (Hamburg: Felix Meiner, 1980), §55, §74. "Der affekt ist Überraschung durch Empfindung, wodurch die Fassung des Gemüts (*animus sui compos*) aufgehoben wird." In Kant's *Kritik der Urteiskraft*, when discussing humor (the joke, and laughter), he writes that "Das Lachen ist ein Affekt aus der plötzlichen Verwandlung einer gespannten Erwartung in nichts." Thus, it is an affect that arises from a sudden transformation of a tension-filled expectation into nothing. Immanuel Kant, *Kritik der Urteilskraft*, ed., Karl Vorländer (Hamburg: Felix Meiner, 1974), §54.

of surprise, but nevertheless presupposes the "rush" of surprise to give *an affect* its suddenness (and open honesty) in contrast to the endurance (and hidden cunningness) of a passion.[18]

We can take this as a leading clue. Even if one wanted to challenge Kant's conclusions, or the more contemporary analyses and conclusions (by, say, Husserl, Merleau-Ponty, and others as well), which suggest that a perception is already affectively significant, we would be hard pressed to assert that surprise is affectively neutral. The fact that surprise is, well, "surprising!" already suggests that it bears some affective resonance in the very experience.[19] In a surprise, we are moved, and moved on the level of "feeling"! Before going further in terms of surprises' status as an emotion or not, let me turn to its valence.

By valence, I understand the experience's positive, negative, or neutral tonality, or in general its "feeling-resonance." For example, when we experience shame, no matter what the shame is about, this emotion is always accompanied or colored by a negative valence. Even if we want to experience shame, it is this negative quality that we would desire. In a similar way, guilt is also qualified by a negative valence; it concerns a negative experience of a transgression of some kind. By contrast, hope is always given with a positive valence, as is loving. Whether or not pride is something we should or should not experience, it is always given with a positive valence. It would be possible to give other examples, but I want to turn to the instance of surprise.

The interesting thing about surprise where its valence is concerned is that its valence cannot be specified in advance. But this is not the same thing as saying that it is neutral. In a surprise, we are moved; a merely rational subject could never experience surprise. In surprise, however, we can be moved in different ways. The experience of surprise can be given with a positive valence (a loved one shows up at my door), or a negative valence (my high school friend took his own life), or an indeterminate valence (my best friend just proposed to my sister, and "I don't know how I feel about that.")

If startle is affectively significant, and only that, how do we characterize surprise? I suggest that surprise belongs to the sphere of emotions (and is not a mere affect). But if there is a more fundamental difference between surprise and startle, it is that surprise counts as an emotion: Surprise is an *originating receiving* as an acceptance of what was not expected—a "receiving" in feeling—which does not

---

[18] See Depraz, "Surprise, Valence, Emotion."

[19] If we examine a discordance of an otherwise concordant experience, say, walking down the stairs while conversing with a friend, a misstep might constitute a *rupture* of that experience of walking, and if I am very much absorbed in the conversation, *I may not even notice* the misstep until after I reflect on what happened—how I scuffed my shoes, for example, or how I hurt my knee. Even if something is placed in my way and I trip, and then I regain my footing, this may be unexpected, it may be a rupture of my gait, but would this necessarily be constituted as a surprise? I don't think so, since simply because things do not go according to how we expect them to run their course does not necessarily mean that it is surprising. That is, we could experience a rupture of experience, and this rupture or discordance (as an anomaly of concordance) could be integrated merely within a passive level of experience.

mean that this is an active decision making process. Surprise is more than an experience of an unexpected givenness, and more than a turning toward an affective allure, and thus more than an affective life-function. Rather, while surprise presupposes a rupture or a disappointed expectation, and is at least a kind of turning toward as an initiation into "activity"; and while a certain coming into relief affectively can be a prerequisite for surprise, the experience of surprise is still something different from that. It is a moved awakening to or turning toward, which is itself *an "original" or "creative" receiving within feeling*, a creative taking up of an affectively significant event, of emergent being or non-being in the unique manner of accepting what I can't accept, a being thrown back on the experience—as joyful, pleasant, hurtful, sad, etc. It is creative in the sense that it is generative "improvisationally" beyond the givens. All this occurs without it being a process that takes place through a judgment or a reflection on the situation.[20]

Thus, surprise belongs to the sphere of the emotions, insofar as it is more than the experience of a mere rupture, disappointment, non-fulfillment, or something unexpected, and as more than a startle-reflex. It is an emotion in part because of the creative way in which we receive the situation in feeling through which we are moved. The fact that this or that could surprise or not suggests that it is operative on the level of the *person* in some sense. But this does not mean that surprise is lived as a moral or interpersonal emotion.[21] In surprise, I am caught off guard and thrown back *on the experience*. I would now like to examine this aspect further by characterizing this folding back on experience as a "disequilibrium" in surprise, and by contrasting it with a "diremptive" experience, which occurs within some (but not all) moral emotions.

4 Disequilibrium and Diremption

I distinguish surprise as an emotion from the field of moral emotions. By moral emotions, I mean those emotions that are essentially interpersonal or that arise in an interpersonal nexus (and not whether they are good or bad).[22] I have distinguished affects from emotions and specifically the moral emotions. Affects are feeling-states and pertain to who we are as psychophysical beings, where we would find experiences like pleasure or pain, being ill at ease, tickling, and arousal. Here is where I place startle. Under "emotions," I include experiences like regret, remorse, fear,

[20] Hence, there is a fundamental difference between "I accept what I cannot accept" and an "I accept that I cannot accept." The former would characterize, in part, surprise; the latter only that my commitments do not entail what I thought they did, or alternately, a second order epistemic affirmation of what is not the case: "I believe that I can't believe that X is the case." The latter would not necessarily be the experience of surprise.

[21] See Anthony J. Steinbock, *Moral Emotions: Reclaiming the Evidence of the Heart* (Evanston, Il: Northwestern University Press, 2014).

[22] See Steinbock, *Moral Emotions*.

longing, and surprise (though this is only a sampling). These are emotions, without being moral emotions, because they can occur without any essential relation to personal "otherness" in that experience. They are genuine emotions, however, because they all presuppose an "order" or even "disorder" of the heart—to use a phrase from Pascal—and are lived in some way toward some other as bearer of value in a "creative" or personal manner.

In order to provide surprise with still sharper contours, I also want to distinguish its internal structure as an emotion from a set of moral emotions. Specifically, I characterize surprise as an *experience of disequilibrium* in distinction to the experience of moral emotions like embarrassment, shame, and guilt and the like, which are always *diremptive experiences*. The difference concerns, in the one case, being thrown back on the experience, and in the other, being thrown back on myself before another.

4.1 Diremptive Experiences

Let me begin with a diremptive experience. A diremptive experience is an experience in which I am given to myself as in tension with a basic sense of myself as before another or others. In short, I am thrown back on myself before others in a way that calls me into question. Diremptive experiences are moral emotions, but not all moral emotions are diremptive experiences. For example, repentance, humility, loving, and trusting are not diremptive experiences, though they are moral emotions.

A diremptive experience can occur either as a mere infraction or rupture (an anomaly) in who I take myself to be, or as a more substantial challenge to who I am. In the former case, we would have an embarrassment, namely, a "mere" discordance of a concordant dynamic orientation of who I am. Let's say that I show up at someone's house and discover that I have holes in my socks. If such a situation yields embarrassment (it need not, of course), it is because it is given as an infraction of my general character (for me) as someone who for instance, takes pride in his socks. However, if it is only an infraction, it is because it is experienced as only an incidental breach. Importantly, however, it is sufficient to throw me back on myself as before another: I am usually well-dressed, but now I have holes in my socks and my friends see it and I see them noticing it. I may apologize or make an excuse or simply turn red: I feel uncomfortable because this has violated in some way a sense of myself as before another.

Guilt, shame, and humiliation are other, more serious kinds of diremptive experiences. Let me focus on shame because of its kinship to embarrassment. While shame is related to embarrassment as a diremptive experience, it is phenomenologically distinct. In this case, the diremption is lived as more than a mere infraction because the event or action threatens a "re-configuration" of my character. On the one hand, there are now two (or at least two) competing ways of being: for example, on the one hand, a well-dressed person, and on the other, someone who cannot take care of himself or who may be "poor" (where poverty is accepted as a communal disvalue). The point here is that rather than the diremption being incidental in relation to who I sense myself to be, I experience it as a reflection on my character. However, even though there are two potentially competing orders or ways of being as a disorientation of my orientation, there is still "only" a threat of complete reorientation. This is because the two ways of being lived at this moment are not given on the same level; one is experienced as more basic than another. Instead, through shame, I am revealed to myself, before others or another, opening up the possibility of self-critique. While shame is always an experience with a negative valence, shame may provide the possibility of a positive reorientation to whom I "truly" am, and thus open the possibility of a genuine self-critique. In short, without both of these: (1) this co-givenness or orientations and (2) without one being more basic, there would be no shame, but merely alternative ways of being; I would not experience any kind of diremption at all.[23]

Now, let's examine the experience of surprise. I go over to a friend's house, take off my shoes, and find that I have a hole in my sock! What happened? I am surprised to find that I have such a hole because these were brand new socks! *If* this is a matter of surprise, I can be surprised without it being before any other, real or imaginary. Further, surprise can take place without my being thrown back on myself at all. Certainly, I could subsequently reflect on what I did, namely, that I put on the wrong socks, that perhaps my shoe rubbed a hole in the sock, and so forth, but this is not essential to the experience of surprise. I could wonder about the being of sockness; I could admire my friend's beautiful new socks; I could pay attention to which socks I put on, etc.

I characterize the surprise experience as a "disequilibrium" because I am "caught off guard," and thrown back *on the experience*, not on myself; I am before the experience, as it were, thrown back on the experience, but not necessarily before others, and not necessarily *thrown back on myself*.[24] Instead, I am disoriented sheerly in relation to my previous orientation. In this way, surprise is neither a diremptive experience nor a moral emotion. In short, surprise does not reveal me to myself as before another, but it does catch me off guard and throw me back on the experience.

[23] See my *Moral Emotions*, chapter 2. Shame and guilt (as well as pride) belong to the order of the emotions of self-givenness. Shame is self-revelatory (as is guilt), whereas surprise is not.

[24] Certainly, I can be thrown back on myself before "another"—where precisely I am that "other," like when I surprise myself (as in the example given above). This means that surprise can be a diremptive experience in certain circumstances. However, it need not be, and this is one reason that, while an emotion, it cannot be understood on the same order of embarrassment, shame, or guilt. I would like to thank Ellie Anderson for bringing to my attention this issue of a possible diremption within the experience of surprise.

4.2 Surprise and Gift

It is commonplace to associate the gift and surprise. Even Jacques Derrida, known for his very careful and subtle analyses, will assert as if it were self-evident that the gift, "if there is any," must harbor the structure of surprise precisely as irruptive and unforeseeable.[25]

An experience of surprise requires a futural orientation in order that the latter be called into question. While a necessary component, however, this is not all that is required for a surprise to emerge. If a surprise arises as something unexpected, it is due in part to the temporal mode of givenness that we can call generally an expectation. In order to delimit the experience of surprise further, I would like to distinguish the latter from the experience of a gift. I do this because we often associate a gift with something surprising. My point here is that whereas *a* surprise (the thing, matter, or event) takes place correlative to the subject's experiencing surprise as an emotion, a gift arises correlative to the experience of humility as a moral emotion.

While loving is an openness to any other—i.e., to all dimensions of being or reality, and in this sense, already "metaphysical"—it is most profoundly a personal movement and thus is given as a moral emotion; humility is the way in which we spontaneously receive ourselves (most profoundly) in loving. Like loving, humility is relational, and reveals me to Myself as not self-grounding.[26] This revelation of Myself in humility ("Myself" as being given by another before another, as relational, as not self-grounding) is expansive, interpersonal, such that I accept myself from another, as accepting the givenness from another and the contributions from others to what I do and to who I am. Thus, even if from the perspective of pride, I "lose" or "forget" myself, in humility, "I" receive myself as a recovery of Myself as relational, since humility is the way I am self-given in the process of accepting or receiving. Unlike pride, humility is the experience of the not self-grounding character of who I am. Further, unlike other emotions of self-givenness, namely, pride, shame, and guilt, however, I am not self-given in a thematic way in humility. And unlike shame and guilt, I am not given to myself in a diremptive fashion. Instead, I am given as having received Myself.

I began with considerations of temporality, especially the mode of expectation, in order to explicate the experience of surprise. For something to be experienced as surprise, we have seen that an expectation must be in play. How is it the case for humility? What kind of temporal modes or modes of temporal givenness are in play? What is given in humility?

It is important to recognize that not every futural opening of an experience is an expectation or founded in one. Hoping, for instance, has a different temporal meaning from an expectation. I do not expect or wait for something when I hope; rather,

---

[25] Jacques Derrida, *Donner le temps. 1. La fausse monnaie* (Paris: Galilée, 1991), 152–57; 186–90; 198. English translation by Peggy Kamuf, Jacques Derrida, *Given Time: 1. Counterfeit Money* (Chicago: University of Chicago Press, 1992), 119–23; 147–50; 156.

[26] For this notion of the "Myself," see my *Moral Emotions*.

hoping is characterized as a sustainable and patient awaiting-enduringly. Further, while I might be disappointed that a hoped for object or event does not arise, I would not be surprised if the so-called impossible did not occur.[27]

The temporal mode of the future as an *expectation* is not in play at all in humility. That is, humility does have temporal meaning, but it has different temporal modes of givenness, which I have called elsewhere a presence-at, a remembering-ahead, and a devotion. That is, since humility (or being humble) is qualified as an openness toward and a reception of what is given, its mode of temporality relative to this openness and reception is a presence-at. This kind of presence-at is reflected in a being thankful for. Second, humility is not only related to the way in which something is accepted or received, but it is also a way in which I receive myself in and through being other-oriented, and the way in which I receive Myself as having been given to myself prior to being able to choose myself. If we understand receiving Myself in this way, then we can speak of a mode of givenness qualified as an "ante-memorial reception." Finally, since presence-at does not motivate anything new to come, it does not anticipate more to be thankful for, and is not lived as an expectation. Nevertheless, there is a futural sense to humility, but this is related to a reception of what is given and of whatever is to come, without it being an anticipation of what is to come. Futurally, it is a kind of "accepting-ahead" without anticipated end. Accepting-ahead is not a mere passive experience, but is lived "actively" as a devotion toward someone or something. Thus, even though humility is not oriented temporally like an expectation, it does have a temporal meaning, namely, accepting-ahead as devotion.[28]

My main point here is that in humility, I do not experience an expectation as one might find where it is an issue of a fulfillment of an intention, an anticipation of a return, or in general what we can call "just deserts." In being humble, I experience that I deserve nothing. To experience that I merit nothing is, however, is *not* based on a depreciation of myself (in which case the focus would still be on myself), but rather in my devotion to some other, as it were, as received. In humility, I have no "just deserts" in the twofold sense that humility is not a matter of justice (as a third party adjudication), and I do not dispose myself to another expecting something in return, with the attitude that I deserve this or that. Thus, there is no disappointment of expectation in humility. In fact, such a reception with no "just deserts," is what qualifies humility. There is no expectation of a fulfillment through what I intend, no disappointment, no rupture. Indeed, I cannot even regard myself as non-deserving in order then to have the experience be one of humility. Experiencing that I am non-deserving, for example, is only possible in and through the reception of the gift in humility; I cannot begin it as believing that "I am not deserving" before anything happens. That would be pride or self-denigration (which is another experience of pride).

---

[27] We could undertake a similar analysis for, e.g., trusting or repentance. See my *Moral Emotions*, esp., ch. 5 regarding the experience of hope.

[28] See my *Moral Emotions*, ch. 7.

In humility, there is no expectation of merit, and there is no expectation of return. Rather, there is nothing to expect. It is not *because of* humility that the gift emerges, but the gift emerges "in" humility. I cannot "intend" to be humble and have the experience of humility.[29] This is why disappointment is not operative in humility as a moral emotion. What is it that we experience in this situation? I suggest that it is not *a* surprise, it is not a rupture of an expectation, it is not the "unexpected," but what is received is received *as* gift. To expect nothing in humility means in the moral sphere that I deserve nothing in particular, which is why everything in humility is received *as gift*.

Likewise, when one receives in humility, what is received is received as a gift; only post facto is it "humbling"; only after the fact do I feel myself as distinctly "undeserving." Again, humility admits of no intention, no fulfillment, no disappointment in the customary perceptual sense. Accordingly, Max Scheler writes that in humility one accepts all things with thanks, from the most subtle pleasure to the grandest bliss; we do this without ever imagining that we deserve even the smallest part.[30] The humble person does not give thought to the proper order of things, but accepts what comes with gratitude and without the thought of merit. Surprise, according to its peculiar structure, in accepting what I cannot accept, makes an initial gesture of rendering something at our disposal. According to its essence, however, the gift is not at my disposal. It is true that the gift is received, but in the moment of its reception, it is not *as* at my disposal. It is the attempt to make the gift as something at my disposal that contributes to its disappearance as gift. It may well be that a gift requires a sacrifice. But if this is the case, then sacrifice (or offering) takes place on the part of myself as receiver, not on the giver. It is not the one who offers who loses something in giving, but the receiver who must have done so such that what is given is received as gift. What has been offered up, as it were, is the prideful self, but this has already been done through the acceptance of the other person. I "receive" Myself, but I do not "find" Myself.

Let's say that out of the blue we receive a love poem from an enemy. On one level, this same matter could be experienced as a surprise and yield an experience of surprise (where it might be a rupture of an expectation, be taken as unexpected, etc.), and on another level, it could be received as gift in the experience of humility, in which there is no expectation functional at all. As a whole, the poem could be experienced as a surprise gift. But, these levels are phenomenologically distinct, and in addition need not be given together. I can experience something as a surprise without it being given as a gift, and I can experience something as a gift, without it being a surprise. Thus, surprise as an emotion is not a moral emotion—as evident in its juxtaposition to humility—and a surprise is not reducible to a gift, and vice versa.[31]

---

[29] See my *Moral Emotions*, comparing pride and humility.

[30] Max Scheler, *Vom Umsturz der Werte: Abhandlungen und Aufsätze*, ed., Maria Scheler (Bern: Francke Verlag, 1955), 18.

[31] It could happen that I epistemically "expect" that someone may thank me for something I did for him (knowing his character), but in fact, I never expect anything at all for what I did, not even any kind of acknowledgment, and in this sense, "expecting thanks" does not even cross my mind.

The problem that gives rise to the identification of surprise and gift originates in part from ignoring the relation between loving and humility as distinctive moral emotions peculiar to persons in which the gift can appear as gift. My "intention" (contra Derrida) is not turned toward the gift—in which the gift becomes the figure of the impossible—because for us it is not a matter of intending a gift. In fact, *it is not about the gift*. Further, humility is not an "intention"—at least in any ordinary epistemic (i.e., perceptual or judicative) sense. There are indeed different modes of givenness.[32] To be sure, simply changing terms (from "intention" to "devotion") is not the point. The point, however, is the interpersonal relation in which the gift arises by not aiming at the gift.[33] It is beyond surprise and non-surprise. The "subject" does indeed not give or receive a gift, because it is not about the subject, and it is not about aiming at a gift; rather, it concerns the interpersonal nexus in which a gift appears as such and can only appear as such.[34]

5 Conclusion

My attempt in this chapter has been to delimit the experience of surprise, and to examine some of its structural features in relation to the epistemic and emotional spheres. I have done this by examining surprise's relation to being as a belief-structure, namely, as an accepting or believing what I cannot believe. I then noted how surprise is a being caught off-guard, and an attentive turning toward, but in such a way that I am thrown back on the experience. Surprise as an emotion is distinct from a startle which is operative as an affect in several respects: startle requires no reconstitution of sense, whereas surprise does, especially as *I am thrown back* on the experience in the latter; startle is always sudden, whereas surprise need not be marked by suddenness; startle can remain within the passive, anonymous sphere, whereas surprise is always operative either within activity or as a transition from the passive to the active levels of awareness; and startle is a reflex, whereas surprise as an experience is an emotion, presupposing an original or "creative" taking up or receiving of a situation.

---

[32] See Anthony J. Steinbock, *Phenomenology and Mysticism: The Verticality of Religious Experience* (Bloomington, IN: Indiana University Press, 2007), "Introduction." And see my *Moral Emotions*, "Introduction."

[33] It is true that receiving the gift can also a transformation of the one who receives, but this already takes place in the presence of another. In pride, there is an exclusion of the gift through a rigidity of the self rather than a "being with" the other person. Receiving the gift, then, is an openness to the other as the openness to myself/Myself, which means the transformation of Myself as open to the other person. It is essentially interpersonal. Ultimately, I receive Myself as from another, as not self-grounding.

[34] Compare, Derrida, *Donner le temps*, 39; Derrida, *Given Time*, 24. See chapter 5.

Yet, although surprise can be considered as an emotion, it is not a moral emotion. The former is marked by an emotional disequilibrium (constituted by being thrown back on an experience), which is distinct from being thrown back on myself before another, or what I call a diremptive experience (such as we find in certain moral emotions, such as embarrassment and shame). Finally, even though we might have something like a "surprise gift," a gift as such is given in and by the moral emotion of humility (which has a different futural temporality than an epistemic expectation), and is distinct from *a* surprise, which is given in the emotional experience of surprise, and which still presupposes the temporal mode of expectation. I conclude that while surprise is neither an affective startle reflex on the one hand, nor an interpersonal moral emotion, on the other, it still can counts as an emotion. Finally, a surprise is essentially distinctive from a gift, which is received in humility.[35]

[35] This chapter was originally published as "Surprise, Gift, Humility," in *It's Not about the Gift: From Givenness to Loving* (London: Rowman & Littlefield, 2018).

Surprise, Valence, Emotion: The Multivectorial Integrative Cardio-Phenomenology of Surprise

Natalie Depraz

Abstract The present paper questions the taken-for-granted identity between surprise and emotion, according to its common meaning in psychology. I therefore hypothetize that the introduction of the concept of valence offers a crucial contribution in order to dismantle the above mentioned identity, thus paving the way for a more fruitful understanding of surprise, namely including cognition, bodily experience, time and language.

Such a discussion about the emotional valence-character of surprise therefore requires to open up the fitting epistemological background for it. Instead of a mere body-brain interaction I will suggest the necessity of a more encompassing setting, namely: body-*heart*-brain, descriptible with the double-face rhythm of surprise, both lived emotional and cardio-vascular: hence the hypothesis of *cardiophenomenology* as an integrated epistemology including heart into the body-brain circulation: far more than the mere one-sided third person neurological level, the heart both in its physiological *and* in its lived affective aspects thus represents a unique possibility of intertwinning first person lived affections and third person organic preconscious sensations.

The hypothesis of cardiophenomenology unfolds the general framework of the inquiry into the dynamics of surprise as *including* an emotional component while *not* being an emotion *as such*. In the end surprise will be identified as an encompassing circular phenomenon.

Keywords Surprise · Emotion · Valence · Heart system · Cardiophenomenology

For the audio-video presentations of the Conference "Surprise, an Emotion?" at the University of Carbondale (25–26 september, 2013), and especially to hear the initial oral presentation, go the following link: www.youtube.com/watch?v=KYanpUeuyeQ.

N. Depraz (✉)
Department of Philosophy, University of Rouen, Mont Saint Aignan, France

N. Depraz, A. J. Steinbock (eds.), *Surprise: An Emotion?*, Contributions To Phenomenology 97, https://doi.org/10.1007/978-3-319-98657-9_2

1 Introduction

"I prepared a surprise for you!" The woman who speaks on the phone with her husband, whose birthday is that very evening, is eager to prepare him for a *good* surprise. Inviting his best friend from home, making a reservation for the concert group *London Grammar* that he is so fond of, booking a weekend holiday in Istanbul, which is the Eastern City he has been dreaming of for years to visit… The husband listening to his wife starts imagining what kind of present it might be: going to the restaurant just at the corner of their home, having sex and watching TV at the same time, discovering the new string his wife bought to "honor" him… He thus feels a genuine pleasure in anticipating the pleasure he will feel that evening, while his wife on her side also feels some true pleasure in anticipating the pleasure he (and she) will have when discovering the content of the surprise she has been preparing for him. Both are looking forward… When coming home, the husband welcomes his best friend and opens the letter containing two tickets for the concert and the promise of an Istanbul weekend. "Are you pleased with your birthday presents and the evening we spent?" his wife asks him later that night after his friend had left. "Well," the husband answers, "it did not exactly meet my expectations! I hoped we would have dinner – the two of us as lovers – and quietly watch TV, have sex and go to sleep early, since we never have the opportunity to do so!" In fact, he was highly disappointed and even a bit angry at her, because she was not even able to anticipate what would really please him: In fact, to him, it was really a *bad* surprise! In turn, his wife will be unpleasantly surprised of course by his unexpected reaction… She was so much looking forward to seeing him happy!

Such a situation, more frequent than we may first think, remarkably illustrates the inherently valence-ladden character of surprise. It is in virtue of the propriety of valence that surprise is not merely equivalent to an event, which may be neutral with regard to valence and not necessarily embedded into a temporal dynamics, which includes the emotional quality of its anticipatory pleasure or tension, its consumatory pleasure (be it satisfaction or disappointment), and of its immediate aftereffects (be they releaving or regretful). In short, a surprise is good or bad, pleasant or painful, or is *not at all* a surprise. An event may be critical, historical, physical, mathematical, but it is not inherently constituted by its valence. One of the main issues of my investigation in this contribution will consist in analyzing and arguing more precisely about such a difference. Another important issue will be to investigate the kind of relationship that is to be found between surprise and emotion (to be distinguished from valence as we will see). From the situated example I detailed above, it is clear that the emotional component obviously appears at various moments and in multiple forms along the different phases of the process of surprise. I aim here at defining more concretely the multifarious modes of emotional appearing within such a process, whereas I will also demonstrate how and why surprise *as such* is *not* an emotion, contrary to some important available stances in philosophy and in psychology. In the end I will hypothetize how the introduction of such a third term into the distinction between surprise and emotion (namely, valence) may well pave the way for a more fruitful understanding of surprise.

Such a discussion about the emotional valence-character of surprise finally requires disclosing the fitting epistemological background for it. Instead of a mere body-brain interaction, I will suggest the necessity of a more encompassing setting, namely: body-*heart*-brain, descriptible with the double-face rhythm of surprise, both lived emotional and cardio-vascular. Hence the hypothesis of *cardiophenomenology* as an integrated epistemology, including heart into the body-brain circulation. Far more than the mere one-sided third person neurological level, the heart both in its physiological *and* in its lived affective aspects represents a unique possibility of intertwinning first person lived affections and third person organic sensations into a unique pre-conscious continuum of experience.[1]

My first step will be to question the obviousness of the taken-for-granted identity between surprise and emotion, according to the question mark of the title of the conference (and present volume). Secondly, I will introduce the concept of valence in order to investigate further the complexity of the relationship between surprise and emotion, suggesting a possible resolution to the difficulty of their connection. Finally, with the hypothesis of cardiophenomenology, I will unfold the general framework of the inquiry into the dynamics of surprise as *including* an emotional component while *not* being an emotion *as such.* Surprise therefore is identified as an encompassing circular phenomenon.

1.1 Surprise and Emotion

Let me start with reviewing the different aspects of the connection between surprise and emotion: their longstanding equivalence in psychology, the thematic absence of surprise within main phenomenologies of emotion, and finally its inclusion into a mere cognitive process.

1.1.1 A Longstanding Taken-for-Granted Contention: Surprise as a Primary Emotion Among Others

For most rationalist modern philosophers as well as for some contemporary leading psychologists, surprise is a specific experience which belongs to a generic form called "primary emotion."

---

[1] An experience therefore is all the more pre-conscious when it is merely lived through. Typically many sensory and emotional lived experiences are pre-conscious in the sense that I am immersed in or permeated by such states, moods or fluctuations without being aware of them. But their are not unconscious because I may become aware in the aftermath of what I lived, if for example a specific event reactivates the previous sensation or emotion. But I would also extend such a pre-conscious lived dynamics of experience to imaginary, perception and remembering insofar as they are lived through in a way Husserl would name « passive » (in the sense of receptive).

1. *The philosophical surprise: a primitive affection*

In *The Passions of the Soul* (1649),[2] Descartes thematizes surprise in an indirect way, his explicit entry into it being "admiration." The latter is defined in the well-known §70 as a "sudden surprise of the soul." Let me quote the thorough definition in French: "L'Admiration est une subite surprise de l'âme, qui fait qu'elle se porte à considérer avec attention les objects qui lui semblent rares et extraordinaires." Descartes provides us here with an articulated description made of five different moments: First, an objet triggers me; second, I judge it as new (rare, extraordinary); third, I feel a sudden surprise; fourth, as an first effect, I feel admiration; fifth, as a second effect, I consider it with attention. Further, this five-sequenced process is announced earlier (§53) as a kind of pre-definition: "Lorsque la première rencontre de quelque objet nous surprend, et que nous le jugeons être nouveau, ou fort différent de ce que nous connaissions auparavant, ou bien de ce que nous supposions qu'il devait être, cela fait que nous l'admirons." Surprise results from the first meeting of an object we evaluate as new and unknown, and its immediate effects are admiration and then attention. Of course, if we limit ourselves to the narrow definition of admiration as a "sudden surprise of the soul," surprise and admiration are synonymous, and we can assert that the Cartesian name of surprise *is* admiration, the latter being the first primary passion without any opposite among the six primitive passions, including love, hate, desire, joy, sadness (§53). However (and we will come back to this point below), upon closer inspection we see that far from being identical, admiration is in fact *the emotional immediate after-effect* of surprise and attention its cognitive after-effect – surprise being itself triggered by a new object that is judged new.

Will we be happier with Kant's *Anthropologie in pragmatischer Hinsicht* (1798)?[3] Contrary to Descartes, Kant provides his reader with quite a sober approach of surprise through the definition of affect: "Affekt ist Überraschung durch Empfindung." First *Überraschung* endows affect with a sensory component; besides, it contains the German root „rasch," which means „quick," thus defining the *temporal* modality—suddenness—of emotion, here as *Affekt*, and it is therefore opposed to *Leidenschaft* (passion), which is characterised by its duration, its evolution in time.

In conclusion, surprise shares three features for both Descartes and Kant: (1) its mode of temporality: suddenness, (2) its primacy among emotions, *and* (3) its generic link but not direct identity with emotion.

2. *The* psychological *surprise: a basic emotion among others*

The question is easier with the recent psychological tradition. Among many others, let me mention here Paul Ekman on the one hand and Robert Plutchik on the other hand. Indeed, both quite pregnantly put surprise to the fore as a basic emotion among other basic emotions.

[2] R. Descartes, *The Passions of the Soul*, Translated by Stephen Voss, Hackett Publishing Company, Inc., 1989.

[3] I. Kant, *Anthropology from a pragmatic point of view*, ed. Robert B. Louden, Introduction by Manfred Kuehn, Cambridge University Press, 2006.

In his article "Constants across cultures in the face and emotion" (1971), co-authored with W. Friesen, P. Ekman (similar to the seventeenth century philosophers) identifies six basic emotions: anger, fear, surprise, disgust, sadness, and enjoyment.[4] He thus provides a partial mapping with Descartes' primary passions, namely, with surprise-admiration, sadness and enjoyment but not with love, hate and desire; instead preferring anger, fear, and disgust. From an empirical study based on facial expressions accross cultures, shown in photographs to several subjects, he concludes that there is a *universality* of such emotions. Regarding surprise more specifically, Ekman finds it is hard to determine what surprise is beyond its physiological expression as an "increased visual field from widened eyes." With such a bodily definition Ekman greatly depends on the criteria Darwin first formulated in *The Expressions of Emotions in Man and Animals* (1872), where surprise is presented either as an adaptive physiological function or an evolved communicative expression, with no clear choice between both. Another interesting notation in Ekman is the empirical difficulty he has distinguishing between fear and surprise for some of the subjects he interviewed, which tends to limit the universality of both emotions.[5]

In subsequent years, Robert Plutchik, as early as 1980, developed what is well-known as a "wheel of emotions" in *Emotion: Theory, research and Experience.*[6] It is made of eight basic emotions (including surprise) and of eight advanced and more complex ones, each resulting in the combination of two basic ones.

The eight basic emotions are presented according to an opposition principle: on the one side, joy, trust, fear, and surprise as "basic emotions;" on the other side, sadness, disgust, anger, and anticipation as "basic *opposite* emotions." For example, disgust and trust are opposites, but since the model is circular they are in fact linked beyond their opposition, which is resorbed because of the sphere-dynamics. Same for surprise and anticipation, apparently opposed, but in fact circularly inner-linked. But what is striking and interesting here is the fact that the opposition principle between both lists is not ruled by a unique exclusive criterium, namely by a positive/negative value-distinction. Indeed, though opposed, fear and anger are both "negative" emotions. Besides, let me notice that neither Descartes nor Ekman, who both provided lists of primary passions/basic emotions including surprise, built the latter on the basis of an exclusive positive/negative value opposition. While love/hate and joy/sadness are such standard value-opposites, admiration-surprise and desire are positively valued and not endowed with any contrary (Descartes). As for Ekman, he is mainly interested in negatively valued emotions: fear, anger, and disgust, which

---

[4] P. Ekman & W. Friesen, "Constants across cultures in the face and emotion" (1971/1993), *Journal of Personality and Social Psychology*, 17, pp. 124–129.

[5] Let me also mention the primacy of fear in William James' account of emotions in the corresponding chapter in *Principles of Psychology*. Here fear clearly appears as a possible name for surprise, which is besides not thematized at all by W. James.

[6] R. Plutchik, *Emotion: Theory, research, and experience: Vol. 1. Theories of emotion*, New York: Academic, 1980, and *Emotions and Life: Perspectives from Psychology, Biology, and Evolution*, Washington, DC: American Psychological, 2002.

have no opposite. As far as surprise, it is the only one positively valued with no opposite, and joy/sadness finally are the only polarized "couple" that share a commonground with Descartes. Thus Plutchik only unfolds in a more systematic way the non-exclusiveness already at work with his two predecessors. He keeps the unique coupled basic emotions joy/sadness; he provides the Ekmanian disgust with a positively valued opposite: trust, which is here a "new" basic emotion; he innovatively "couples" fear and anger which are two negatively valued emotions both absent in Descartes and both newly provided but uncoupled by Ekman; finally, he originally "couples" surprise and anticipation, which are both positively valued emotions, the latter sharing with *trust* the "newness" character of the basic emotion. The question is: What are the *other* criteria that couple fear/anger on the one side, surprise/anticipation on the other side, since they are obviouly not coupled according to the positive/negative value opposition?

While both are endowed with a negative value, fear and anger share a valence-dynamic based on the motor-action inter-active opposition of defense (non-action: fear) and attack (aggressive re-action: anger); as for surprise and anticipation, both enter into a dynamic rather than into an opposition (whatever it may be: negation or reaction). What is its criterium? Their connection is a time-connection. Surprise and anticipation are temporally-embedded.[7] Moreover, they are situated on the same "line" of the wheel, with distraction and amazement on the one side and vigilance and interest on the other side, which well demonstrates how surprise and attention (here as vigilance, interest and distraction) go hand in hand, revealing the intrinsic—also etymological—link between awaiting ('*attent*e' in French) and *attent*ion.[8] Surprise is clearly a process, a duration,[9] thus not reducible to a sudden punctual moment, a shock, as Descartes first thought. Hence, the fact that anticipation understood in a pre-conscious, unwillful, implicit way (in Husserlian terms "protention"

[7] On such a time-embeddedness between surprise and anticipation, see my book *Lucidité du corps*. De l'empirisme transcendantal en phénoménologie, Den Haag, Kluwer, 2001, Section II; see also N. Depraz and F. Varela, "Au cœur du temps: l'auto-antécédance II" in: *Intellektica* n°36/37 (J.-L. Petit éd.), Compiègne, février 2004, pp. 182–205. See more recently my research hypothesis within the ANR I am leading, in "Shock, twofold dynamics, cascade: three signatures of surprise. The micro-time of the surprised body" in: *Surprise at the intersection of phenomenology and linguistics*, N. Depraz & A. Celle eds., Boston-Amsterdam, Benjamins Press, submitted [from my talk entitled "Experiential phenomenology of surprise: 'l'éclair *me* dure'", Proceedings of "La surprise à la croisée de la phénoménologie, de la psychiatrie et de la pragmatique," 21–22 March 2013, University of Rouen/Ens-Cnrs-archives-Husserl. video recording available at http://www.univ-rouen.fr/audio/2010/index.php?vid=337], and Th. Desmidt, M. Lemoine, C. Belzung & N. Depraz, "The temporal dynamics of emotional emergence," *Phenomenology and the Cognitive Sciences*, 2014. Available at the following link: http://link.springer.com/article/10.1007/s11097-014-9377-8.

[8] On such a dynamic between surprise and attention, see my article "An Experiential Phenomenology of Novelty: The Dynamic Antinomy of Attention and Surprise," in *Journal for Constructivist Foundations, Special Issue Neurophenomenology*, Vol. 8, n°3, S. Stuart & M. Beaton eds., 2013.

[9] See N. Depraz, "La surprise. Une dynamique circulaire de verbalisation multivectorielle," in *La surprise à l'épreuve du langage et des langues*, N. Depraz & Cl. Serban eds., Paris, Hermann, 2015.

as a quite primitive organic tension) is intimately linked to the possibility of genuine suddenness. In other terms, there would not be any surprise without a specific quality of anticipation. Interestingly enough Plutchik grasped such an original connection between surprise and anticipation in his own way, even though there is no direct historical connection between his work, Husserl's, Ricœur's and ours.

Furthermore, we have eight advanced and more complex emotions (love, submission, awe-anxiety, disapproval-disappointment, remorse, contempt, aggresiveness, optimism), each resulting in the combination of couples from more basic ones. For example, contempt is the result of anger and disgust, love results from joy and trust. If we stick to the specific case of surprise, anxiety is presented as being the combination of surprise and fear (hence the close link between them), and disappointment the result of surprise and sadness.[10] Such integrated emotions show that we have to deal here with a three-dimensional dynamic model and not with a linear list of emotions opposed one to the other.

To come back to my initial example, the husband and wife both have expectations (even though contrasted) about what the birthday is going to be like (a feast with the husband's best friend, a tv-dinner as lovers). They both build imaginary projections, thus unfolding an anticipatory process, and such anticipations (whatever they are truly) *create* the very possibility of their being surprised – here in a negative way, in the form of a disappointment that is the combined result of surprise and sadness, also mixed with anger in the case of the husband.

So here again, still more obviously than for philosophers like Descartes and Kant, for these psychologists, surprise results in an emotion as such.

1.1.2 The Absence of Surprise in Founding Phenomenologies of Emotion: A Counter-Hypothesis

As a contrast, surprisingly enough, surprise nearly does not appear at all as a theme in major phenomenologies of emotion. The leading question is then: Why doesn't surprise belong to affective acts for phenomenologists or why doesn't it play a role in the phenomenon of affectivity as such? For different reasons indeed, we will see that phenomenologists who attribute an important or crucial role to affectivity did not think it necessary to include surprise in their analysis.

1. *The Husserlian case study*

Even though the founder of phenomenology did not build a full-fledged phenomenology of emotion (and perhaps maybe thanks to it), it is interesting to notice how surprise—though non-thematically—may be caught at different unexpected places and levels of Husserl's analyses.

---

[10] Let me note that Husserl thematizes disappointment as a non-fulfillment of expectations, thus identifying a possible negative name of surprise, which is quite coherent with Plutchik's combined naming of sadness and surprise as disappointment. I will come back to the Husserlian surprise as disappointment below.

Let me insist on three different layers: (a) In the framework of basic bodily tendencies and drives,[11] surprise occurs non-thematically as a bodily reaction of resistance (*Widerstand*) to an alluring novelty and my bodily inner tension undergoes a felt-effect of release; as such, it is quite a primitive pre-conscious organic process; (b) At a higher conscious level, in the course of a time-embbeded perceptual activity (visual most frequently), my being affected by an unexpected object, situation, or person, which occurs in contrast or discrepancy with my ongoing expectations, amounts to a surprise; as a direct effect of my affection, it is named by Husserl as a dis-appointment (*Entäuschung*). For example: I am awaiting a friend and he finally does not come up.[12] (c) At a still more complex level, a gnoseological-interest level, within the dynamics of attention (*Bemerken*), I am striken by something I did not notice first. Surprise appears as a twofold phenomenon, both objective and subjective, or rather intrinsically ruled by noetic-noematic dynamics: 1. what strikes (*spannt*) and becomes salient (*abgehoben*) on the 'objective' side, and 2. a releasing just after-phase (*Entspannung*) on the 'subjective' side, which may then retain an emotional component which is not present in the striking objective aspect of surprise. For example: I am affected by the strident noise of the train that triggers my noticing it.[13] The subjective moment of release may be positive or negative, either in the sense of being angered (I am disturbed in my reading) or excited (I was waiting for my lover coming on train).

At each of these levels of description, be they pre-conscious bodily-organic, lived conscious perceptual, or gnoseological-attentional, the time-process horizon of undetermined awaiting is at work, endowed with a more or less gradual ability from the side of the subject to anticipate what is going on and what will go on. So even though Husserl never thematizes surprise, there are many analyses where a phenomenology of surprise may be sketched out.[14] As a conclusion, surprise can be

[11] Exemplarily in the *Studien zur Struktur des Bewußtseins* that are currently in the process of being edited at the Husserl-Archives in Leuven by U. Melle and Th. Vongehr, namely the second volume dealing with affective acts. In this second volume, a manuscript entitled *Gefühl* (*cerca* 180 pages) was dealt with at the Husserl-Archives in Paris during the year 2014–2015, read through and worked out as a French translation in the framework of a workshop led by N. Depraz and M. Gyemant. I thank the Husserl-Archives in Leuven for permitting us to do so. This manuscript is accepted with the editor Vrin and currently in course of edition.

[12] See my footnote 10 for dis-appointment as an indirect negative polarized surprise of what does not meet my expectations. For this perceptual level of analysis, see *Experience and Judgment* (1939), Northwestern University Press, Paperback, 1975, §21, and more thoroughly, *Analyses Concerning Passive and Active Synthesis. Lectures on transcendental logic* (1918-1926) (1966), American translation by A. Steinbock, Kluwer Academic Publishers, 2001.

[13] See E. Husserl, *Wahrnehmung und Aufmerksamkeit* (1883–1904/1905), *Hua* XXXVIII, 2004, french translation by N. Depraz, Paris, Vrin, 2008, for example Appendix II, or again *Analyses Concerning Passive and Active Synthesis, op. cit.*, §35.

[14] See N. Depraz, "L'inscription de la surprise dans la phénoménologie des émotions de E. Husserl," first given as the opening talk of the monthly Research Seminar "Emotions and volitions" (04/10/2013) at the Husserl-Archives (ENS-CNRS) organized in the framework of the ANR Research project I am directing *Emphiline-EMCO "La surprise au sein de la spontanéité des émotions: un vecteur de cognition élargie"* (2012–2015), published in *Alter* n°23 (2016), « La Surprise » (N. Depraz ed.) under the title "Husserl et la surprise". It is available in its original initial PDF draft at the following ANR link: http://129.199.13.46/spip.php?article492. It was also published in

defined as the contrast inherent in a lived bodily-cognitive time-dynamic phenomenal process (*Spannung/Entspannung*).

2. *Contrasted post-Husserlian Views: scheler, Heidegger, Sartre*

I will go more quickly now into some post-Husserlian perspectives, even though they would obviously require a similarly close look as the one I just provided for Husserl. Specialists in Scheler, Heidegger, and Sartre would equally find places where surprise indirectly appears, in the same way as I was able to detect some in Husserl, thus locating surprise even if it is not thematized.

The point I want to make here is the following: In these phenomenologies where emotion in the global sense of affectivity is clearly and crucially at work (in a manifest way in Scheler,[15] but also in the early Sartre,[16] and still in Heidegger's analysis of *Stimmungen*[17]) surprise needs to be reconstructed.

In Scheler's systematic analysis of the emotional life, affectivity is from the start situated at the social-intersubjective, historical, and moral level. Along with sympathy as the core of all other emotional drives and with love the utmost expression of pure sympathy, Scheler deals with many other moral, other-related affects: shame, guilt, hope, suffering, despair, and so on. In short, Scheler's interest is greatly driven towards high ethical issues (exemplarily the value of human life). With such an extended social embedded framework it is easy to understand why surprise as a lived bodily physiological reaction does not have any clear place.

Secondly, Heidegger's ontology of event (*Ereignis*) may appear as an obvious place to do justice to surprise, at least as a break within time-continuity. The author of *Sein und Zeit* truly brings about a renewed conception of time ruled by the open future-oriented and affectively anguish-ladden indeterminacy of my being-for-death, and a full-fledged analysis of temporally embedded affective dispositions. However, surprise is a far more complex experience, mainly because of its emotional, bodily, and intersubjective intrinsic parameters, and is therefore not reducible to its event-character; as such, it is neutral, lacking any bodily setting and self-orientation. Surprisingly enough, Heidegger's discontinuous anthropology leads him to understand surprise as limited to animals, thus excluding it from humanity. In his *Grundprobleme der Metaphysik* (1929) as well as in *Grundfragen* (1937–1938), the animal is defined by its *Benommenheit* (stupor) and as being in a constant attitude of *Einnommenheit* ("emprise" in French), which is in contrast with

Spanish in the Proceedings of the Conference in phenomenology organized by Leonardo Verano in Columbia in November 2013 in *Eidos,* Revista de filosofia, n°21, Universidad del Norte, 2014 under the title "La inscripción de la sorpresa en la fenomenología de las emociones de Edmund Husserl."

[15] M. Scheler, *Wesen und Formen der Sympathie* (1916), BoD – Books on Demand, 2013; *The Nature of Sympathy*, Routledge & K. Paul, 1954, reimpressed 2008 by Translation Publishers, New Brunswick.

[16] J.-P. Sartre, *Esquisse d'une théorie des émotions* (1938), Paris, Hermann, 2010; *Sketch for a theory of the emotions*, Methuen & Co edition, 1977, Routledge Classics, 2001.

[17] M. Heidegger, *Sein und Zeit* (1927), Tübingen, Niemeyer, 2001; *Being and Time*, Harper Perennial Modern Classics, 2008.

any questioning attitude, and as such is reserved to the human being, the only one who is able to be astonished ("étonné"). In short, Heidegger's understanding of surprise is highly axiological. He therefore "misses" surprise, either confining it to the mere bodily animal stupor or opening it up to humans as a spiritual awakening mind and a questioning astonishment (*thaumas*), which, as we know, is synonymous with the emergence of philosophy itself.

As far as the Sartrian entry into surprise is concerned, it would actually need a closer investigation, insofar as the different places where surprise occurs are multifarious and …unexpected. The first and most obvious analysis to refer to is the one of emotion as a magical consciousness and conduct, developed in the already mentioned *Esquisse d'une théorie des émotions*. Though it is not thematized as an emotion and is in fact not an emotion as such, surprise shares with it a "magical character" as one fundamental way of being opposed to its pragmatic-rational way. Emotion (and surprise) truly corresponds to something magical, that is, an irreal transformation of any event and situation. It therefore refers to a true "magical instant."[18] In *Being and Nothingness*, surprise therefore unthematically comes up as the ordinary reality of contingency; not in the formal logical Leibnizian meaning of what could have been otherwise or different, as a classic positive mode of possibility, but as the very process of *néantisation* [annihilation/becoming nothingness]. Throughout the book, at least three exemplary figures of such a contingency appear: the Other (though in an ambivalent way), the object, and death.[19] Finally surprise refers to daily encounters with others that are highly theatricalized in *Les Mots*. Sartre's autobiography is full of creative and enjoyable situations of surprise that built the very vitality of his life as a child and changed what could have been dramatic events into funny and marvelous situations, into gifts. "J'adorais les surprises. Il y en avait tout le temps chez nous. Cachotteries amusées ou vertueuses, cadeaux inattendus, révélations théâtrales suivies d'embrassements. C'était le ton de notre vie."[20] The author is here full of humour, of light irony about himself, and surprise

---

[18] See P. Ricœur, his analysis of the « émotion-surprise » and his reference to Sartre in *Philosophie de la volonté* I, Paris, Cerf, 1949, p. 238; see also Ph. Cabestan's talk, 13th of June 2013 at the ENS in the framework of a Conference organized by N. Depraz, "La surprise au risque de sa valence émotionnelle," about "Sartre: surprise et magie de la surprise" (http://savoirs.ens.fr/expose.php?id=1375), published in *Alter* (2016), "La surprise" (N. Depraz ed.).

[19] J.-P. Sartre, *L'être et le néant*, Paris, Gallimard, 1945; *Being and Nothingness*, Washington Square Press (1993), Paperback, 2003: about the Other, p. 170: "the malice of others always surprises me…"; but p. 755: "the Other cannot surprise me; the being which is wishing to bring into the world, which is myself-for-the-Other,—this being I already enjoy possessing"; about the object, p. 219: "in the knowledge of a transcendent object indeed there is a revelation of the object, and the object revealed can deceive or surprise us; and finally, about death, p. 686: "but the unique quality of death is the fact that it can always before the end surprise those who wait for it at such and such a date."

[20] J.-P. Sartre, *Les mots*, Paris, Gallimard, Folio Poche, 1977 (1ère éd.), pp. 88–89. See also the famous setting at the hairdresser's: "Un jour, j'avais sept ans, mon grand-père n'y tint plus: il me prit par la main, annonçant qu'il m'emmenait en promenade. Mais, à peine avions-nous tourné le coin de la rue, il me poussa chez le coiffeur en me disant: 'Nous allons faire une surprise à ta mère'. (…) Le surlendemain de l'opération, Auguste était venu voir mon grand-père. 'Je vais, lui avait-il

is omnipresent in quite a developed vocabulary: "cachotteries, révélations, inattendu, théâtre." The natural spatio-temporal determinacy therefore becomes relative, even contingent, and in the end, light and ludical.

In short, our four phenomenologists all seem to neglect surprise as a primarily individual organic short-timing shock that might be shared as an experiential continuum by humans and animals. However, whereas Husserl's unique micro-descriptive interest in the fine-grained inner lived bodily flowing life of the subject allows him to make crucial room for surprise as a micro-rupture, neither Scheler, nor Heidegger, nor Sartre provide an adjusted and fully-fledged phenomenological framework for it. In sum, they primarily stress what I would call relational emotions, that is, either intersubjective or social-ethical (Scheler: affective participation and contagion), existential and enduring affections (Heidegger: anguish, boredom, loneliness) or situational-theatrical (Sartre: horror, admiration, joy).

Of course, if I considered surprise as brought about by a present I give to somebody else, like in the situation I described at the beginning of this article, where the wife surprises her husband for his birthday, then surprise would clearly be a loving, relational phenomenon, even though it fails meeting the husband's expectations and brings about a feeling of disappointment mixed with sadness and anger. However, are we to consider suprise as a moral emotion in the sense of guilt or shame (Scheler), as ontological in the sense of anguish (Heidegger), or as magical and determinacy-free (Sartre)? Even the positive or negative value attached to it (for example disappointment vs. satisfaction) does not seem to be enough to call it a moral, ontological, or irreal-magical emotion.

In conclusion, for each of our phenomenologists, surprise needs to be reconstructed though other themes and concepts. It is not a direct topic; nor is it described as such. Interestingly enough phenomenologists provide macro-structural concepts where surprise may then be further explored as a specific embodied phenomenon.

1.1.3 Shifting from an Emotional Component to a Cognitive Process: A Common Agreement from Ancient and Modern Philosophers

If surprise does not belong to emotionally affective acts, does it mean that we have to refer it to a cognitive process? Such a contention is shared by philosophers coming from quite different contexts; hence its potential relevance. In this respect,

dit, t'annoncer une bonne nouvelle'. Karl fut trompé par l'affable solennité de cette voix: 'Tu te remaries ! – Non, répondit mon oncle en souriant, mais tout s'est très bien passé. – Quoi, tout?' Etc., etc. Bref les coups de théâtre faisaient mon petit ordinaire et je regardai avec bienveillance mes boucles rouler le long de la serviette blanche qui me serrait le cou et tomber sur le plancher, inexplicablement ternies; je revins glorieux et tondu. Il y eut des cris mais pas d'embrassements et ma mère s'enferma dans sa chambre pour pleurer: on avait troqué sa fillette contre un garçonnet. Il y avait pis: tant qu'elles voltigeaient autour de mes oreilles, mes belles anglaises lui avaient permis de refuser l'évidence de ma laideur. Déjà, pourtant, mon œil droit entrait dans le crépuscule. Il fallut qu'elle s'avouât la vérité. Mon grand-père semblait lui-même tout interdit: on lui avait confié sa petite merveille, il avait rendu un crapaud; c'était saper à la base ses futurs émerveillements."

Aristotle's stance in *Poetics* is inaugural: Whereas *thaumaston* (astonishment)—from the Greek verb *thaumadzein*—refers to the "objective" triggering-source of knowledge, *ekplêktikon* corresponds to the immediate subjective effect produced by *thaumaston* on the audience, wherein you can read the Greek root *plêgê* (stroke).[21] These are the two correlative cognitive names for surprise: what triggers my interest as an object-like source of surprise and what subjectively strikes me as an effect.[22] Much later on, among empirists like A. Smith (eighteenth century),[23] pragmatists like C. S. Peirce (nineteenth century),[24] phenomenologists (Husserl), French philosophers (H. Bergson),[25] and then among recent philosophers of mind like D. Davidson[26] and D. Dennett,[27] surprise will be considered a unique "cognitive emotion," with surprise knowledge or epistemic prediction being put in the wrong. Being surprised then amounts to becoming aware of the discrepancy[28] or the mis-

---

[21] Aristotle, *Poetics*, edited and translated by St. Halliwell, (Loeb Classical Library), Harvard 1995, 24, 1460a12-14 et 25, 1460b25.

[22] About surprise in Aristotle, see A. Hourcade, "La surprise dans la *Poétique* d'Aristote" in N. Depraz & Cl. Serban eds., *La surprise à l'épreuve du langage et des langues*, Paris, Hermann, 2015.

[23] A. Smith, "Wonder, Surprise, and Admiration" one feels when contemplating the physical World (1795), Lecture on Astronomy, Section *I: Of the Effect of Unexpectedness, or of Surprise, Glasgow Edition of the Works and Correspondence Vol. 3 Essays on Philosophical Subjects* [1795], ed. W. P. D. Wightman and J. C. Bryce, Indianapolis: Liberty Fund, 1982: "We are surprised at those things which *we have seen often*, but which we *least of all expected to meet with in the place where we find them*; we are surprised at the sudden appearance of a friend, *whom we have seen a thousand times*, but whom we did not imagine we were to see then." Thus surprise is not identified with the rareness or the extraordinariness of its object like it is the case for wonder, but *via* the relational situation of a "displacement" within the most familiar and sedimented context of experience.

[24] Ch. S. Peirce, « On phenomenology » (Harvard Conference 1903), in *Pragmatisme et pragmaticisme* (*OC* I, Paris, Cerf, 2002, pp. 295–296): the definition of surprise was early attested by the founder of pragmaticism, Ch. S. Peirce, who really appears as being *the* philosopher of surprise, while defining the « action of experience as a series of surprises », stresses straightaway its self-generating process.

[25] See for example H. Bergson, *Essai sur les données immédiates de la conscience*, Paris, P.U.F., Quadrige, 2011, p. 56, footnote 1. In Bergson's work surprise is omnipresent as a structural "unexpected rupture" *grounded* in our habits and mechanical routines, for which the philosopher provides numerous figures: the surprise of laughter, of insight, etc.

[26] D. Davidson, (1982) *Rational animals. Dialectica* 36:318–327, and (2004) *Problems of rationality*, University Press, Oxford, p. 7, who retrospectively concurs with Husserl and Smith, while defining (1982) surprise as a proof of the mastery of concepts of true and false belief. For him you cannot be surprised *without possessing some beliefs*. Conversely, if you possess some beliefs you are faced with the possibility of being surprised, because something can happen that may change your mind and your beliefs. Davidson gives as an interesting daily example the case of somebody putting her hand in her pocket and finding a coin. If she is surprised upon finding the coin, then she realizes that her previous belief about her pocket was false. She therefore can be credited with the belief that there is an objective reality independently from (previous) beliefs.

[27] D. Dennett, (2001) "Surprise, surprise," Commentary on O'Regan and Noe. *Behav Brain Sci* 24(5), p. 982: "Surprise is only possible when it upsets belief."

[28] See for example A. Ortony, G. Clore & A. Collins, *The Cognitive Structure of Emotions*, Cambridge, Cambridge University Press, 1988; R. Reisenzein, Meyer, W.-U., & Schützwohl, A.

match[29] between what I thought, believed, or predicted and what actually is. It reveals the non-coincidence between two states of knowledge and indicates the pragmatic necessity of adjusting and coping.[30]

1.2 *The Multivectorial Integrative Dynamic of Surprise*

After reviewing three main contentions about surprise in multifarious scientific and philosophical contexts (that is: (1) its taken-for-grantedness as an emotion in psychology and in classical rationalist philosophy, (2) its quasi-absence in phenomenologies of emotions and finally (3) its interpretation as a (meta-)cognitive emotion-feeling in cognitive psychology, philosophy of mind, and neuro-biology) I would like to argue for my own proposition about surprise, that is, the complex situation of the emotional dimension within the rhythm of surprise as process. My leading contention states as follows: Surprise involves an emotional *and* cognitive component but results in a more *encompassing and integrative circular (time, bodily, expressive-descriptive)* phenomenon. My approach is integrative insofar as I

(1996). "Reactions to surprising events: A paradigm for emotion research." In N. Frijda (Ed.), *Proceedings of the 9th Conference of the International Society for Research on Emotions* (pp. 292–296). Toronto, Canada: IS, and about surprise as a "metacognitive feeling," see R. Reisenzein, "The Subjective experience of surprise" in H. Bless and J.-P. Forgas eds., *The message within. The role of subjective experience in social cognition and behavior*, Philadelphia, PA: Psychology Press, 2000, pp. 262–279.

[29] A. K. Seth (2013), "Interoceptive inference, emotion, and the embodied self," *Trends in cognitive sciences, 17* (11), pp. 565–573: the model suggests that emotion is induced by a mismatch between interoception (understood as a key-bottom-up process of emotional emergence) and the top-down process of prediction of what interoception may be. In that regard, the insular cortex is a key structure involved in the process of comparison between interoception and prediction. When this comparison results in a mismatch, it generates a prediction error signal from the insula to the neural structures involved in autonomic and motor regulations, what eventually trigger reflexes related to emotion. Such a strictly third person computational model describes a neurocognitive process of predictive coding that can be understood as a "neural correlate of protention," based on the Bayesian probabilities approach. It is a brain-centered model that pays no attention to the first person approach toward emotion with no integrative proposal for the experience of emotion. Although interoceptive mismatch can be considered an aspect of surprise, it is a strictly cognitive truth-valued account of surprise. From a broader phenomenological perspective, surprise results greater than the mere moment of the mismatch, as it structurally encompasses the phase of anticipation and aftermath in a global experience involving the three systems of the brain, the body/heart and the context. Since Dennett and Davidson's belief-truth oriented understanding of surprise make a similar extension of surprise to its 'prior' expectation-belief phase, I aim at extending their simple truth-valued conceptions by discussing the broader emotional and heart-bodily levels and by including the aftermath, which is not discussed by these authors. More on this in Th. Desmidt, M. Lemoine, C. Belzung & N. Depraz, "The temporal dynamics of emotional emergence," *Phenomenology and the Cognitive Science*, 2014, *op. cit.*

[30] For more on surprise as an emotionally cognitive process, see N. Depraz, "The surprise of non-sense," in Massimiliano Cappuccio & Tom Froese (eds.), *Enactive Cognition at the Edge of Sense Making,* Palgrave MacMillan, 2014.

define surprise on the basis of five main structural invariants (dynamics of time, enactive lived body, expressive and descriptive language, cognitive process, emotional affect) and of five variables still in exploration (moral, imagination, intersubjectivity-alterity, volition, passivity, pathology). These eleven invariants and variables are extracted from crossing and comparatively co-generating a priori/experimental third-person categories and experiential first-person ones that emerge from microphenomeological explicitation interviews.[31]

As far as the five invariants are concerned, there is a double move at work in the integrative construction of the phenomenon. The first move refers to a negative deconstructing process, in virtue of which I open up the possibility of another meaning for surprise than the common standard one:

Negative Cartography of Surprise

- Not an instant-shock
- Not a physiological startle
- Not a mere description
- Not a cognitive emotion
- Not a primary emotion

The second move reconquers these invariants in a positive way at an upper dynamic-circular level and shows surprise as a multivectorial integrated process:

Integrative Cartography of Surprise

- Duration
- Embodies becoming aware as a recursive "cascade"
- Aftermath indirect verbalisations
- Attentional awaiting and attentional resonance
- Emotional immediate association

1.2.1 The Time-Dynamics: A Structural Condition

Surprise is commonly equated with an event that is said to be "unexpected" or "counter-expected" and would merely amount to an instantaneous shock extracted from any kind of temporal continuity or process. My hypothesis, based on crossed third- and first-person validations, is the following: Surprise is experienced as a break in the time-flow continuity, and such a break is intrinsically linked to the

[31] The explicitation microphenomenological interviews proceed from two different but correlated experimentations: a psycho-linguistic task and a psycho-physiological experiment. 50 interviews with English students are presently completed for the linguistic task (80 are programmed); 45 interviews with subjects with three different groups (depressive, in remission, test-subjects) are now completed (on the basis of 75 subjects for the experiment). Although the corpus is still in the course of being analyzed, some interviews are already transcribed and analysed and allow for preliminary indications which question and/or confirm the a priori categories on the basis of emerging experimental and experiential ones. I will give some examples of such a crossing-checking categorization in the following.

subject's experiential past retentional and future protentional immediate horizon. It is therefore defined as a process rather than as a discrete event, this process taking the form of a circular micro-time dynamics unfolded into subphases. Thus the time of crisis, i.e. discontinuity, is directly jointed to its implicit protentional anticipation and its subsequent remanent resonance.

As third-person philosophical-phenomenological validations, Husserl exemplarily defines the "living present" as experientially structured by an awaiting horizon and a protentional dynamics, where surprise emerges in the form of disappointment, a rupture in the flowing perceptive continuity.[32] Hence the pre-conscious bodily and inner anticipatory tension co-inherent in such a rupture. For Paul Ricœur, surprise is not reducible to a reflex-stimulus but entails an inner duration, which is the very experience of becoming aware.[33]

As correlative first-person preliminary indications, let me mention three different rythms linked to some first person descriptions available in the psycho-linguistic task: (1) One student confronted with Soutine's painting "Carcass of beef" feels "disgust" and says, "Chicken!" to himself. Here surprise is first bodily-emotional and then innerwardly expressed in the form of an exclamation. It is not an instantaneous shock but it is an unfolded, inner duration; even though short, it is at least bi-modal, gustative, and innerwardly exclamative.[34] (2) Another student at the appearing of Louise Bourgeois' "Spider"[35] immediately feels satisfaction, and then, discovering it is only a sculpture, she becomes disappointed. Here surprise is clearly

---

[32] E. Husserl, *Erfahrung und Urteil*, Akademia Verlagsbuchhandlung, Prag, 1939, pp. 94–95; *Experience and Judgment* [1939], Churchill, J. S., and Ameriks, K., translators [1973]; London: Routledge, Northwestern University Press, Paperback, 1975, §21, a) "Der Ursprung der Negation": "(...) statt der Erfüllung der Erwartungsintentionen tritt *Enttäuschung* ein. Z. B. es sei eine gleichmässig rote Kugel gesehen; eine Strecke lang ist der Wahrnehmungsverlauf so abgeflossen, daß diese Auffassung sich einstimmig erfüllt. Aber nun zeigt sich im Fortgang des Wahrnehmens allmählich ein Teil der zuvor unsichtig gewesenen Rückseite, und entgegen der ursprünglichen Vorzeichnung, die da lautet 'gleichmässig rot, gleichmässig kugelförmig', tritt das die Erwartung enttäuschende Bewußtsein des Anders auf: 'nicht rot, sondern grün', 'nicht kugelig, sondern eingebeult'."

[33] P. Ricœur, *Philosophie de la volonté* I, Paris, Cerf, 1949, p. 239: "(...) Comment un bref jugement de nouveauté peut-il avoir pour corps un battement de cœur, une inhibition diffuse, une certaine stupeur qui fige le visage et dispose les parties mobiles des sens à l'accueil ? Pourquoi en retour cette disposition du corps est-elle une disposition de l'esprit à considérer l'objet et à s'attarder sur lui (...) Dès lors la pensée incarnée n'est plus jamais punctiforme ni réduite à glisser infiniment sur les choses sans s'y arrêter; le corps empêche que la rencontre avec le nouveau reste une touche fugitive; il fait que la conscience s'étale, s'écrase en quelque sorte (...) le corps amplifie et magnifie l'instant du penser, en lui donnant pour épaisseur de durée le temps de saisissement du corps; par la surprise une pensée s'impose en quelque sorte physiquement." (My underlining.) See N. Depraz, "La surprise. Une dynamique circulaire de verbalisation multivectorielle," in *La surprise dans le langage et dans les langages*, N. Depraz & Cl. Serban eds., Paris, Hermann, 2015. English trans. by S. Davidson, "Surprise. A Circular Dynamic of Multi-Directional Verbalization", in: *Journal of French and Francophone Philosophy—Revue de la philosophie française et de langue française*, Vol XXVI, No 1 (2018) pp 1–17.

[34] See Transcription 007 Emphiline philo mp3.

[35] See Transcription 010 Emphiline philo mp3.

bi-valenced according to a micro-segment of time. I call such a surprise a "double release." (3) A third student looking at Gustave Doré's "Vision du Paradis"[36] was directly and quasi-physically attracted by the tunnel-effect of the image that she almost dug into it; then discovered two small shadows at the forefront that she did not initially see. She finally glanced at the angels all around the sun. Each time she felt a specific surprise, either visio-kinetic, or focal, or again peripheric, inherently articulated according to a generative "cascade" movement.[37]

These three different first-person experiences of duration help concretize the philosophical hypothesis of duration and also confirm its validity.

1.2.2 The Enactive Lived Body: An Embodied Becoming Aware

For most scientists, startle is the common physiological naming of surprise. Even though many first-person accounts in the microphenomenological explicitation interviews I led also mention an initial startling experience, some do not, and the startling component, though important, is not the whole of the very *bodily* surprise. In this respect, empirist philosophers like Adam Smith and Charles Sanders Peirce define surprise as a process, either intrinsically linked to a context (Smith) or unfolding as a series (Peirce). For Smith, surprise is the contrast between a familiar situation and an occurrence that does not fit in it; for Peirce, surprise is not a punctual occurrence but refers to a series of micro-surprises, so many tiny interruptions in the very process of experience. Surprise and experience therefore are co-extensive.

On the side of first-person descriptions, the testimony of the third student I mentioned above looking at Gustave Doré's "Vision du Paradis"[38] is quite emblematic of the irreducibility of surprise to a punctual startle. The embodied-generative cascade-like surprise she underwent is made of an initial startle-like physical attraction, directly followed by a focal visual awareness and then a peripheric open glancing. The bodily dynamic moves according to the perceptual processs of becoming aware therefore creates a kind of gradual awakening.

Once again, the experiential first-person indication helps concretize the structural philosophical statement of a contextual-serial process of becoming surprised.

1.2.3 Language, Cognition and Emotion

Let me proceed more quickly for the three other invariants of surprise, since they map the time and body invariants with their own content. With regard to language, we need to take into account the whole variety of expressions of surprise, from the bodily language made of patterns such as exclamative, verbless interrogative

[36] See Transcription 004 Emphiline philo mp3.

[37] For more details about such the processual duration of surprise, see N. Depraz, "Shock, twofold dynamics, cascade: three signatures of surprise. The micro-time of the surprised body", *art. cit.*

[38] See Transcription 004 Emphiline philo mp3.

sentences and interjections to aftermath verbalisations (be they narrations or rationalisations), which often use the verb "to be surprised": "It was surprising because...," or "I was surprised by the red colour... ."[39] As far as cognition is concerned, a common scheme is shared by Husserl and Davidson despite their opposed philosophical anchorage. The familiar sedimentation/belief is the experiential dynamics enabling the perceptive contrast corresponding to surprise.[40] As for affect, our results lead us to make the claim that surprise is not an emotion in the sense of a basic feeling like fear, anger, disgust, joy or sadness, contrary to common statements in the literature. Although surprise remains frequently associated with emotional valence when correlated to joy, grief, anger, or pleasure, it may also appear as a neutral, mixed, or epistemic emotion, i.e. as a violated expectation that affects both action and cognitive processing.

1.3 Suprise and Valence

Along the detailed examination of the situation of the emotional dimension in the process of surprise, I already came to mention the valence-ladden character of surprise, which is first of all linked to its polar emotional contents (e.g. anxiety/hope, horror/marvel, disappointment/satisfaction).

1.3.1 Emotional Blank and Modulations During the Process of Surprise

So let me detail now more precisely how emotion is involved all along the process of surprise without surprise be an emotion as such.

I mentioned above (and analyzed elsewhere) the rhythmic, dynamic model of surprise made of three temporal micro-phases.[41] At each of these micro-phases a specific quality of emotion accompagnies the process. With phase 1, the awaiting phase, there is a generic quality of tension (*Spannung*) linked to the protention opening, which, depending on the kind of event in different emotional perduring qualities, may be specified for example either as hope or as anxiety. With phase 2, the crisis phase, which is the core of the dynamic of surprise, the emotional intense and shock-quality occurs just after the visual perceptive impact of the image and sometimes after the explicit bodily startling reaction and it is described for example either as "horrible" or as "marvelous." Hence the emotional blank at the very appearing of

---

[39] See N. Depraz, "La surprise. Une dynamique circulaire de verbalisation multivectorielle," *art. cit.*

[40] See N. Depraz, "The surprise of non-sense," *art. cit.*

[41] For a philosophical-phenomenological account, see N. Depraz, *Lucidité du corps. De l'empirisme transcendental en phénoménologie,* Den Haag, Kluwer, 2001, IIème Section, and more recently and scientifically atunned, see Th. Desmidt, M. Lemoine, C. Belzung & N. Depraz, "The temporal dynamics of emotional emergence," *Phenomenology and the Cognitive Science, art. cit.*

the image. With phase 3, the aftermath-effect phase, like in phase 1, we have to do with perduring emotional qualities which accompany the process of release-resonance, either as disappointment versus satisfaction, depending on the content of the event, if the latter does not affect any longer, or with a perduring emotional tension of horror or marvel if the image remains in my mind after its physical disappearance. In short, at each of the micro-phases of the process of surprise, the emotional dimension occurs with different contents and qualities, either as perduring or as shock-like. It regularly accompanies each phase and often immediately follows the triggering perceptive-bodily reaction in the form of an associated dimension, *but it does not seem to initiate the process of surprise in neither of its phases.*

To sum up, we have the following scheme:

1. The awaiting phase: perduring-accompanying emotion as *tension* (in terms of valence: hope vs. anxiety)
2. The aftermath phase: perduring-accompanying emotion as *release* (in terms of valence: satisfaction vs. disappointment)
3. Question: Is there an emotional component during the crisis phase and, if yes, when?

There is an emotional blank at the very appearing of the stimulus. Emotion appears just before it (protention) and occurs just after it as an immediate aftermath-effect (in terms of valence: neutrality). There is also a primacy of bodily movements (cardiac and peripheral) and of body language (exclamation, interjection). In conclusion, I offer a multivectorial hypothesis about surprise. There is no reason to abstractly give a primacy to emotion in order to characterize surprise, even though emotion is truly associated along the whole process in different forms.

1.4 Valence at the Crossroads of Neuro-Physiology, psychology and Lived Ethics

Actually, valence already indirectly operates in my discussion of the complex articulation between surprise and emotion.

The goal of this step is precisely to put valence to the fore. Why? I choose here to introduce valence as an exploratory concept characterized by its two-fold polar dynamic (-/+),[42] with the following hypothesis: Valence characterizes more precisely the affective dynamic of surprise rather than emotion as such, which always remains a partial and static state, at least if it is understood as a constituted emotion, which it is most often.

[42] For a first investigation of valence with this hypothesis, see N. Depraz & F. Varela, 1999/2004/2011: "At the source of time. Valence and the constitutional dynamics of affect," 1999, PURH, Ipseity and Alterity (S. Gallagher and Ph. Brun eds.) *Ipseity and Alterity: Interdisciplinary Approaches to Intersubjectivity,* S. Gallagher & S. Watson eds., Rouen, Presses Universitaires de Rouen et du Havre, 2004; E. Thompson ed. *Emotion, Journal of Consciousness Studies,* 2011.

1. *Origin of the term and further use in psychology of emotions*

The first occurrence of the term is due to Kurt Lewin, a German-American social psychologist particularly active in the twenties-thirties. He defended his PhD under Carl Stumpf's supervision, was first involved in schools of behavioral psychology, and then changed his orientation and undertook work with psychologists of the *Gestalt* school of psychology including Wertheimer and Köhler. In this context, he founded the principle of force-field analysis, which provides the framework for investigating the factors (forces) that influence a social situation. These forces are primarily of two opposite kinds, either driving movements toward a goal (helping forces) or blocking movements (hindering forces).[43] The term valence, which characterizes such a double-opposed movement, entered English usage in psychology with the translation of works of Kurt Lewin from German in 1935. In his book *A Dynamic Theory of Personality*,[44] Lewin characterizes valence as a dynamic of opposite forces, which he basically names attraction and repulsion. Such conflictual dynamics is meant to foster the learning of the individual in a context of action, where he develops his abilities while coping with such conflictual dynamics.

Beyond its initial action-learning embedded meaning, it is considered as an operative concept to define the very structural dynamics of emotions in psychology. More particularly, it means the intrinsic attractiveness (positive valence) or aversiveness (negative valence) of an event, object, or situation.[45] By extension, it is also used to characterize and categorize specific emotions: negative ones (anger, fear) have a negative valence while joy has a positive valence. The term is also used about the hedonic tone of feelings, affect, certain behaviors (for example, approach and avoidance), goal attainment or nonattainment, and conformity with or violation of norms. In that respect, ambivalence can be viewed as a conflict between positive and negative valence-carriers.

As a conclusion, two different meanings of valence emerge. Either it is a dynamic of opposed values, or it is the value (positive, negative) of a specific emotion. That is, either it is a global concept eventually applicable to emotions but not characterizing them intrinsically, or it is an invariant feature of emotions. We will see how surprise will receive a contrasted understanding depending on the chosen meanings and articulation between valence and emotion.

2. *Controversial statements about the relation between valence, surprise and emotion*

If valence is considered as one of the invariant features of emotions insofar as it is the value (positive, negative) of a specific emotion (for example, sadness as a negative value or trust as a positive value), then surprise may be characterized as

[43] K. Lewin, "Defining the 'Field at a Given Time'," *Psychological Review.* 50, pp. 292–310, 1943, repub. in *Resolving Social Conflicts & Field Theory in Social Science,* Washington, D.C.: American Psychological Association, 1997.

[44] K. Lewin, *A Dynamic Theory of Personality.* Selected Papers, New York-London, Mc Grawhill Co., 1935, English trans. by D. K. Adams & K. E. Zener.

[45] N. H. Frijda, *The Emotions.* Cambridge University Press, 1986. p. 207.

deprived of valence since it is not exclusively linked to a positive or a negative value but its very dynamics encompasses the global process itself of valuation. In that respect, it is not an emotion attached to a value, be it positive (like joy) or negative (like anger), and it will be described as a physiological startle associated with an intense cognitive state of rupture of time-continuity triggering prediction-error. As a consequence, surprise will be completely disconnected from valence as a neutral cognitive and bodily process.[46]

If, counter-hypothesis, valence is seen as *the very dynamic* of opposed values, that is, as a global concept eventually applicable to emotions but not characterizing them intrinsically, then it may share such a global dynamic with surprise, both being structured as a polarity. In that respect, surprise can have any valence, that is, it can be neutral/moderate, pleasant, or unpleasant. Valence is therefore inherent in surprise but not in emotion.

Whereas the first two options (strictly cognitive, or merely emotional, defining surprise as one of the basic emotions as we say first) remain one-sided, only the third option is in my view complex enough to articulate surprise to valence as its immediate associate without reducing it to valence itself, so that the core of surprise may remain non-exclusively invaded by emotion.[47]

3. *The present hypothesis: an encompassing operative dynamic concept*

Hence my understanding of valence as a structural polar transversal encompassing-dynamics starting with physiology (bodily, behavior) and reaching through ethics (norms, valuing, ambivalence) *via* its association with constituted emotions.

(a) *Neurology*

Let us keep in mind our multivectorial model of surprise made of three microphases and its subsequent understanding as a process including the instantaneous crisis, its implicit anticipatory time, and its immediate perseverative aftermath. As we already mentioned, valence operates at each phase as a twofold tension (anxiety/hope), shock (horror/marvel) and resonance (ill at ease/relief).

At the neurological level, a number of scientific works have established the dominant role of the amygdala in order to stock, remember and modulate the valence (positive or negative) of a sensory event.[48] Such a value seems to be time-embedded insofar as the event which is value-ladden as pleasant/unpleasant is situated within

[46] For a provisional sketch about this assumption, see R. Casati & E. Pasquinelli, « How can you be surprised? The case for volatile expectations », *Phenomenology and the Cognitive Sciences,* 2007, 6: 171–183, p. 172; about surprise as a metacognitive emotion, see A. Ortony, Clore, G., Collins, A., *The cognitive structure of emotions, op.cit.*; for the cognitive interpretation of surprise as a prediction error within apredictive coding, see A. K. Seth, *art. cit.*

[47] For a more detailed account about these statements, see N. Depraz, "The surprise of non-sense," *art. cit.*

[48] J. J. Paton et col., « The primate amygdala represents positive and negative value of visual stimuli during learning », *Nature*, 2006, 439, 7078, p. 865–870.

its anticipatory dynamic and its resonance.[49] As such, a value is also attached to each phase. We therefore have to distinguish between an anticipatory pleasure, a consumatory pleasure, and a perseverative pleasure, so much as between an anticipatory anxiety, a consumatory deception, and a perseverative disgust. More concretely, the value at work at each phase may well be contrasted. For example, the husband looking forward to spending a great evening with his wife (anticipatory pleasure) but ends up disappointed (consumatory displeasure); or the wife who is anxious to please her husband (anticipatory displeasure) and is then happy to spend a nice evening with him (consumatory pleasure), but sad to realize thereafter his disappointment (perseverative displeasure).

(b) *Ethics*

What is experienced at a visceral bodily level through pleasure or displeasure and results reflected in the limbic brain-system has strong correlates at the level of moral emotions. For example, the wife's preservative displeasure may be named after a moral valence an emotion of guilt, whereas the husband who is looking forward to spending a nice evening with his wife will feel a moral emotion of hope. The happiness of the wife during the birthday-evening can be translated as pride from a moral point of view, while the husband's consumatory displeasure, triggering disappointment, may end up in a moral feeling of contempt against the person who did not guess his true desire. In short, from visceral bodily valence to ethical valuing via psychological valuation there is a continuum of different levels of experience and approach, which receive multifarious emotional names.

(c) *Pheno-psycho-physiology*

We already mentioned that emotional quality appears sooner in the form of a tension linked to awaiting and later in the form of a release or a perseveration in the aftermath. During the crisis-phase, bodily startle, exclamative language and cognitive becoming-aware are dominant. At the very beginning of the emergence of the stimulus, visual focussing may open the way for an intense emotion of horror, or for a motor-startle, which is then directly associated with an emotional shock. At the very beginning, surprise appears to be attentional-cognitive, which correspond on the physiological level with a heart-deceleration, then followed up with the emotional intensity of the shock or the horror, which gives way to a heart-acceleration.[50]

---

[49] E. A. Murray, "The amygdala, reward and emotion," *Trends Cogn. Sci.*, 2007, 11, 11, pp. 489–497; L. Pessoa, "On the relationship between emotion and cognition," *Nat. Rev. Neurosci.*, 2008, 9, 2, pp. 148–158.

[50] For a more detailed analysis based on two patients, for which we correlatively analysed the explicitation microphenomenological interviews and the heart frequency dynamics during the crisis-phase (6 seconds), see N. Depraz & Th. Desmidt, "Cardio-phénoménologie," Strasbourg Conference (2014, April), *La naturalisation de la phénoménologie vingt ans plus tard*, J.-L. Petit and T. Pozzo eds., *Cahiers philosophiques de Strasbourg*, n°38, 2015.

Here, then, is my hypothesis as a starting point to be confirmed by first person indications: Valence is a generic dynamic category involving a twofold mode of emotional appearing at specific moments of the integrated multimodal process of surprise and along a neuro-physiological-ethical continuum.

(d) *The Bielefelder school in the late nineties: a pioneering integrated psychological model of surprise*

In order to clarify more precisely the specificity of the above model, I would like to situate it in the framework of a model of surprise that was built nearly twenties years ago in Germany thanks to a group of psychologists around Wulf-Uwe Meyer and including Rainer Reisenzein and A. Schützwohl.

One pioneering article was published as early as 1996 under the following title: "Analyse von Reaktionen auf überraschende Ereignisse: Ein Paradigma für die Emotionsforschung."[51] One remarkable commonpoint of this "process" model and mine is its integrative multimodal dimension. Contrary to physiological, psychological, linguistic or cognitivist approaches, it does not merely define surprise as a startle, as a basic emotion, as an expression, or as an upsetting of my previous beliefs, but it takes into account these different parameters into a unified view. This is exactly my aim and my claim, and it seems to me (and this is my specific point) that phenomenology enables such a synthetic integration.

In that respect, the first moment in the article[52] sums up quite well both the integrative methodology and the central (paradigmatic) role of surprise for a theory of emotions. Let me quote it a bit extensively:

> (1) surprise has a characteristic phenomenological quality that can vary in intensity; (2) surprise is, at least in the standard case, intentional or object-directed (that is, when one is surprised, one is normally surprised about something); (3) surprise is, at least in the standard case, caused by cognitive appraisal processes (typically thought to consist in the appraisal of unexpectedness); (4) surprise appears to be associated with a characteristic nonverbal expression (Darwin, 1872/1965) and more or less typical physiological reactions (presumably those also characteristic for the orienting response to novel stimuli); (5) surprise is probably of old phylogenetic origin (as suggested by the existence of similar reactions in other species); and related to that, (6) it has a plausible function that represents a solution to a recurrent adaptive problem (for more detail, see below and Meyer, Niepel, & Schützwohl, 1994). Furthermore, in the case of surprise answers to central questions of emotion psychology may be more readily forthcoming than has been the case for other emotions, for the study of surprise offers a number of methodological advantages: (a) Surprise can be easily and reliably induced in the laboratory with excellent control of its temporal onset and intensity; (b) apart from subjective experience assessed by verbal report, several objective indicators of surprise are available (interruption or delay of ongoing action; nonverbal expression; physiological responses); and (c) ethical problems can be largely avoided.

[51] R. Reisenzein, Meyer, W.-U., & Schützwohl, A., "Analyse von Reaktionen auf überraschende Ereignisse: Ein Paradigma für die Emotionsforschung," in H. Mandl (Hrsg.), *Bericht iiber den 40. Kongreß der DGPs in München 1996,* pp. 830–836, Göttingen, Hogrefe, published in English under the title: "Reactions to surprising events. A Paradigm for Emotion Research," in N. Frijda (ed.), *Proceedings of the 9th Conference of the International Society for Research on Emotions,* Toronto, Canada, 1997, pp. 292–296.

[52] *Art. cit.*, p. 292.

So Reisenzein and his colleagues seem to take into account multifarious components (cognitive, physiological, evolutionary), but also, interestingly, a phenomenological component, which is besides mentioned first. Of course, it is limited to the intensity and the object-directedness problems, but the mention of a phenomenological approach is rare enough to be a good indication of the extensiveness of the model. Besides, it seems to be that this article opens up a multimodal approach that is not really reflected in the articles refered to in the bibliography (above all Meyer's), which are mainly restricted to a cognitive and psycho-evolutionary approach,[53] as if it promised more than what was actually established at the time.

The very mention of the expression of "subjective experience" in another article signed in 2000 by Rainer Reisenzein alone and included into a volume centrally dealing with the theme of the "within"[54] is telling about the unique broader scope of the author[55] and especially his stress on surprise with a "phenomenal character" and an "immediate awareness," that is "pre-reflectively" known.[56] Nevertheless, the way surprise as a "subjective" experience is included into a broadly cognitive process and a psycho-evolutionary model as a "de surcroît" component that fits well in it remains in my view insufficient. Of course, it is an important step forward to take consciousness into account in the framework of a standard third-personal, neutral cognitive process as it seems to be the case with Reisenzein's colleague Meyer. However, to understand surprise as a mere "conscious state" or as having a "conscious aspect" is not enough.[57] Why? Because consciousness is here only added in a pre-existing third-person model as being the explanatory cause of surprise. Instead, my idea is that (1) surprise is not the causal result of consciousness, but, on the contrary, it is the very experiential lever for becoming aware, understood as an emergent-dynamics. Therefore it is as such an experiential process. (2) Rather than considering consciousness as a structural third-person component to be added "from

[53] R. Reisenzein & Ritter, D. (1999), *Surprise: A "metacognitive feeling."* Manuscript under review; T. Ruffman & Keenan, T. R. (1996), "The belief-based emotion of surprise: The case for a lag in understanding relative to false belief," *Developmental Psychology,* 32, 40/19 ; A. Schützwohl (1998), "Surprise and schema strength," *Journal of Experimental Psychology: Learning, Memory, and Cognition, 24,* pp. 1182–1199; Schützwohl, A. (1999), *Surprise: A psychoevolutionary paradigm for emotion research,* New York: Springer; A. Schützwohl (1999), *The structure of the intensity of surprise.* Unpublished manuscript, University of Bielefeld; A. Schützwohl & Reisenzein, R. (1999), "Children's and adult's reactions to a schema-discrepant event: A developmental analysis of surprise," *International Journal for Behavioral Development, 23,* pp. 37- ; J. Stiensmeier-Pelster, Martini, A., & Reisenzein, R. (1995), "The role of surprise in the attribution process," *Cognition and Emotion*, 9.

[54] R. Reisenzein, "The subjective experience of surprise." In H. Bless & J.-P. Forgas eds., *The message within: The role of subjective experience in social cognition and behaviour*, Philadelphia, PA: Psychology Press, Chapter 15, pp. 262–279.

[55] R. Reisenzein, *art. cit.,* p. 262: "When describing the model depicted in Figure 15.1, I made no explicit assumptions about the consciousness versus unconsciousness of the processes that it postulates. However, as noted in the introduction, surprise does have a conscious aspect that probably serves specific adaptive functions, and the theory of surprise must pay heed to this fact."

[56] R. Reisenzein, *art. cit.,* p. 262.

[57] R. Reisenzein, *art. cit.,* p. 266.

outside," it is in my view more fruitful to consider it "from within" as the very experiential first-person dynamics of surprise. It means involving a strict first-person methodology in the form of Vermersch' explicitation microphenomenological interviews, and not only third-person self-reports in the form of a quiz as does Reisenzein. Moreover, it is interesting to notice that such a third-person process model of surprise, though subjective, does not take time into account as a crucial sequential inner component of the dynamics, protentional-anticipative and remanent-resonant; neither does it mention any moral-ethical aspect of it, which is included in the integrated model of surprise I promote.[58]

1.5 Towards Cardiophenomenology[59]

I aim here as a provisional final step to delineate the methodolological and epistemological approach that in my judgment brings about the most fitting and encompassing for the multivectorial model of surprise I presented. Of course, cardio-phenomenology as a general approach is also meant to be used and checked beyond the phenomenon of surprise, and it is hoped that it will thus contribute to its validity. However, since cardio-phenomenology emerged within the very exploration of surprise as a generic-dynamics situated beyond (or beneath) the distinction between emotion and cognition, it is also meant to provide such a phenomenon with a paradigmatic role in the understanding of experience as such.

---

[58] Let's notice that the collective experimental work set out in the nineties does not seem to have been carried on, except for R. Reisenzein's own individual work, who is now fullprofessor at the University of Greifswald and mentions among his fields of research "the psychology of surprise." Among more recent publications, see two 2009–2010 articles entitled "Surprise" (in Ramachandran's *Encyclopedia* and in *Oxford Companion to the affective sciences*); see also with Schützwohl, 2012: "Facial expressions highly surprising event."

[59] "Cardio-phenomenology" as the integrated approach corresponding to the model of surprise unfolded in this article will be only sketched here. It was first mentioned as "a heart-centered model" in "The Rainbow of emotions. At the crossroads of neurobiology and phenomenology," 2008, *Continental Philosophy Review* 41, (B. Heiner ed.), pp. 237–259, as a direct following up at the Collège de France in a Conference org. by B. Andrieu & A. Berthoz, "Le corps en acte" (22/23 sept. 2008) with a talk entitled "Le cœur: corps du corps," then in Tours, 2009, 23th-24th June, at the Journées *Psychiatre, Phénoménologie, Philosophie de l'esprit*, (Th. Desmidt, org.) on the theme "Emotions et psychiatre," with a talk entitled again "Le cœur: 'corps du corps'." It was again more thoroughly presented explicitly under the label "cardiophenomenology" in Berlin, October 2013, in the framework of the *Mind and Life Conference* "European Conference on Personal and Societal Change from the Contemplative Perspective" (10–13 October) with a talk entitled "On becoming surprised: an experiential cardio-phenomenology of depression" (unpubl. Powerpoint available). The full-fledged theoretical contention of cardio-phenomenology is defended in N. Depraz & Th. Desmidt, "Cardiophénoménologie" in: *La naturalisation de la phénoménologie, 20 ans après* (Colloque International, Université de Strasbourg, 22–23 avril 2014), J.-P. Petit éd., *Cahiers philosophiques de Strasbourg*, 2015. "Cardiophenomenology: a refinement of neurophenomenology" is actually submitted to the Journal *Phenomenology and the Cognitive Sciences*.

So why cardio-phenomenology? First of all, what does it mean? The prefix *cardio* refers here to the ancient greek word *καρδία*,[60] which means the heart, both in the physiological meaning of the pumping muscular organ[61] and in the current, sometimes metaphor-like but also psychological and philosophical understanding of the hearth of affect or more broadly as the core of affectivity.[62] In English, as in Greek and in French (*cœur*), one unique word encompasses the polar continumm which goes from the organic-physiological (embryogenetic, neuronal, cellular, and muscular) to the lived affective bodily dimensions, thus accounting for the unity of the person as she unfolds into both a first, lived affective person and a third, objectified organic person. In German, by a contrast, two different words are available to describe such a subjective-objective polarity of the living being, be it at the body or at the heart level. I speak of the fold *Leib/Körper* and analogically *Gemüt/Herz* when accounting for the continuous bodily or heart-dynamics.[63]

Thus, in resonance with such preliminary language-anchored indications, I claim with "cardio-phenomenology" that a heart-centered model is a most integrative epistemological view. In order to argue for such a strong contention, two steps are required, which will help frame (and re-frame) our spontaneously brain-centered, oriented-tendency model. On the basis of such an epistemological re-framing, it will be possible to unfold the requirements proper of cardio-phenomenology as a refinement of neurophenomenology, which I will only sketch here.

What are these two initial methodological steps? Firstly, I wish to present some experiential-methodological and ontological arguments that show how the heart-

---

[60] A. Bailly, *Abrégé du dictionnaire grec-français* (1895), Paris, Hachette, 1901, article καρδία, cœur: 1. *(Anatomie)* organe du corps. Ex: "Πάσχω τὴν καρδίαν," Je souffre du cœur; 2. siège des passions ou des facultés de l'âme, de l'intelligence. Ex: "ἐκ βάθους καρδίας," du fond du cœur.

[61] Very early for example: J. A. E. Eyster & E. C. Swarthout, "Experimental determination of the influence of abnormal cardiac rhythms of the mecanical effificency of the heart," *Arch Intern Med (Chic)*. 1920; 25(3):317–324. See also Gesell R. A. "On relation of pulse pressure to renal function," *Am J Physiol*, 1913; 32:70. For a phenomenological inspired approach, see Merleau-Ponty's *La nature* (1956–1960), Paris, Seuil, 1995, pp. 188–199, which relies on G. E. Coghill's dynamic anatomy and morphogenesis, *Anatomy and the Problem of Behaviour*, NY/London, Mac Millan, 1929, as well as on A. Gesell & C. B. Amatruda's *The embryology of behaviour* (1945), New York, Harper, 1982.

[62] About the heart and its intelligence proper, see very early B. Pascal's "Le cœur a ses raisons que la raison ne connaît point" in *Pensées*, Section IV Des moyens de croire, n° 277. See also a remarkable early intuition in P. Ricœur's *Philosophie de la volonté. 2. Finitude et culpabilité*, Paris, Seuil, coll. Points, 1960, 2009, Chapter 4 "La fragilité affective," pp. 125–126: "le cœur serait le moment fragile par excellence (...) une philosophie du cœur est-elle possible, qui ne soit pas rechute au pathétique (...)?" See finally N. Depraz, "Délimitations de l'émotion. Approche d'une phénoménologie du cœur," *Alter*, n°7 Emotion et affectivité, 1999, pp. 121–148, in particular, p. 143 onwards, entitled: "La singularité cardiale de la mobilité émotionnelle."

[63] About the german word *Gemüt*, see Kant's *Anthropology* (1798), §87, Husserl's *Studien zur Struktur des Bewußtseins*, still unpublished manuscript, in course of translation into French by N. Depraz & M. Gyemant; G. Strasser, *Das Gemüt*, Utrecht, Uitgererij Het Spectrum, 1956, English transl. *Phenomenology of Feeling*. An Essay on the Phenomenon of the Heart by R. E. Wood, Pittsburgh, Duquesne University Press, 1977; Ricœur, *op. cit.*, p. 125 and for an equivalence with *thumos*, pp. 153–183, Depraz, *art. cit.*, p. 124.

system is an integrative system, which helps recasting the explanatory gap; secondly, I will make a constructive proposal about the unique dynamic circular living rythmic of such a system.

1.6 The Integrative Heart Centered System

Dualism (whether ontological or more softly methodological) haunts longstanding debates in Cartesian-inspired philosophy but also in more recent cognitive sciences. In that respect, the initial mind-body problem is being replayed in the consciousness/brain issue, with a radical monist alternative, the only virtue of which being its abstract reductionism.

We took the first step toward re-casting the problem by turning to emergent theories, which consider conscious states and brain-dynamics as irreducible and co-emerging properties, as it is claimed by F. Varela's neurophenomenology. In connection with emergentism, enactive theories also extend cognition to the bodily environment, thus taking into account a structural autopoetic interaction between the living being and its living world.

1. *A first-person phenomenal remedy to the explanatory gap*

Putting forward a heart-centered system amounts to integrating such recent emergent-enactive thrusts into a still more concretely phenomenal-oriented dynamics of the living. Why? Because the heart-hypothesis appears to be favored both on the empirical and on the phenomenological levels. Like the body, the heart is a unique reality, both physio-organic and innerly lived. More than the body, in other terms being "a body of the body," its core lets it emerge affectivity as the root of cognition, thus bringing about en enlarged and renewed cognition. In short, the heart becomes the matrix of the person as both lived (affection) and organic (muscle), or again, the core of the weaving between the first- and the third-person experience of the subject.

How is it? The heart-dynamics reveals an unheard phenomenal fold-continuity of the living being, which undoes the remnant abstract statements about duality or monism of mind/brain. Indeed, whereas you will never be able to feel your neurons in your brain, whatever advanced perceptions you cultivate, you may quite easily and spontaneously feel your heart beat! Such an immediate observation of my and your heartbeats as being directly "self-feelable" is a daily common shared experience and provides us with a robust experiential clue: I feel my pulse; I feel my heart beat in my breast; my pulse races when I get anxious or after running; I feel my pulse reduce at rest; the rhythm of my heart always beats faster (and I can feel it) when I have drunk too much coffee or fallen in love etc…

2. *An embryogenetic argument*

Besides such an experiential argument, we have an embryogenetic one: The heart self-organizes as soon as the third week of the embryo develops spontaneous contractions independently of the brain, and even before the brain is functional. In short, the birth or the functioning of the heart, quite differently from other organs, does not depend first on stimulations from the brain-system.

What embryogenesis tells us consequently is that there is a real – even though relative – self-independance of the heart, with the brain's control upon the heart occuring later, after three months of life. In summary, at an early primary bodily growth stage, the heart-system furnishes the living being with the most radical integrative option, also integrating the nervous and brain system, including a control function.[64]

1.7 The Cardiac Rythmicity as an Interactive Circular Dynamics

Now, less as a critical argument than as a constructive proposal, I will present cardiac rythmicity as a clue to sew the discrepancy of the explanatory gap generated by the irreducible ontological discontinuity between mind/consciousness *and* brain.

First let us notice the homology between the neuro-organic and the cardio-vascular dynamics. In the same way that the brain- and body-networks inter-act together and with the close environment *via* the afferent-efferent nervous circuit, the heart as a pump generates a double integrated co-generative circulation of blood within the whole body and of air thanks to the breath-rhythm as an interaction between inside and outside.

On such a generic dynamic, we can distinguish between specific transversal interfaces of the heart-rhythmic that interrupt the still dual distinction between the organic and the lived, that is, the empirical and the transcendantal.

1. The circular organicity of cardiac pulsation

Such a cardiac pulsation operates at different integrated levels: (a) a physiological pendular spontaneous movement initiates the contraction and dilatation of the heart and activates the blood circulation from center to periphery; (b) the phenomenal cardiac pulsation functions in me in a pre-conscious automatic way, but I may easily consciously access it while taking the pulse (mine or the other's); (c) the speed of the cardiac rythm (acceleration/deceleration) structurally correlates with emotional peaks (anxiety, joy) and decreases (calm, apathia), so that we have an observable unified continuity between the physiological and lived appearing of the heart; (d) the open directedness of desire and appetite opens up the way for a self-determination of the subject as inhabited by the free rhythm of his or her pulsion without any object.

2. The structural inter-active breath-rythm

As a common basis of blood-circulation and of breath-interaction, the heart system takes together the double intra- and inter-organic rythmic move of the body.

[64]About such a view, see M. Merleau-Ponty, *op. cit.*, along with Coghill's and Gesell's contentions.

The circular blood dynamics thus reverberates the inter-active and subjective dynamics self-other: While breathing in, I welcome and adopt alterity in myself; while breathing out, I expulse and give something of myself outside to the other. Both blood and breath dynamics therefore co-emerge and co-express each other as an organic phenomenal manifestation of the relational intra-/inter-subjective dynamics of inner-alterity and self-other inter-actions.

3. The Altering Self-Mobility Micro-fluctuations

At the subpersonal, neural level, the limbic system generates an inner self-regulation: the thymic system (from the greek *thumos*) thus increases its ability to self-immunity, not by resisting outside agressions but by welcoming them as components of its bodily integrity.

At the inter-personal level of our affective-ethical relationships, the subjective self appears as an individuated unity inherently inhabited by relational abilities; these are then less defensive in case of a relational crisis than they are expressive of the latter and involve a therapeutic dynamic.

1.7.1 Cardio-Phenomenology: Neurophenomenology's Next Step!

A theoretical move from a brain-centered to a heart-centered model finally has its adequate methodological counterpart in cardiophenomenology.

1. The Productive Limitations of Neurophenomenology

Francisco Varela's 1996 pioneering thrust with neurophenomenology put to the fore mutual generative constraints between the time-embedded subjective experience and the subpersonal neuro-dynamics. It primarily concerns the interplay between a priori lived phenomenological categories and subpersonal, neuro-dynamics phase-categories. A limitation of such generative constraints lies in the huge time discrepancy between a priori, *timeless* philosophical categories (here the Husselian living present as a protention/impression/retention structural dynamics) and the subpersonal, *milliseconds* neurodynamics. One major bias with neurophenomenology (despite of and because of its pioneer-like thrust) concerns the irreducible ontological discontinuity of each dynamics, which prevents it from having any *operational* articulation. If we look more closely at the respective dynamics, either a priori or subpersonal, we must in fact deal with two third-person approaches.

Cardiophenomenology by contrast suggests experiential mediations along the organic-lived fold in between a priori transcendantal and neuro-dynamic categories, so that a phenomenal continuity may appear and operative correlations emerge. The challenge therefore is to apply instead generative constraints to a twofold continuous unitary pre-conscious experience, organic (heartbeating) *and* lived (emotional), in order to fill in the gap even before creating it, while also starting from an experiential, bi-dimensional fold rather than from an ontological irreducible distinction.

2. The relevance of cardio-phenomenology: a promising research program

Given the experiential organic-lived graduation at stake, more appropriate tools may be used to provide cardiac measures and micro-features proceeding from microphenomenological explicitation interviews sharing a common seconds-scaled time, both objective (physiological-cardiac) and lived (perceptible). We thus gain refined mutual generative constraints operating at the crossroads of a genuine first-person level and of a physiological third-person level, so that we may succeed in synchronizing the timing of the organic *and* the lived affective experience of a both startling/lived surprise into a common seconds-scale.

Hence some crucial pre-requisits of cardio-phenomenology, in order to make available a refined descriptive phenomenology of surprise as a multivectorial dynamics including valence at each of its three micro-phases:

1. Physiological measures[65] are needed to get relevant descriptive third-person indications about the bodily-cognitive process of surprise as startle. Psycho-physiology thus appears as a new complementary field besides neurodynamics, founded (1) on a less subtle micro-timing than the neurological one, but more easily mappable with the lived subjective grosser timing, (2) a direct access to a first-person experience where I can spontaneously feel my heart beating (contrary to my neurons).
2. An experiential (and not only theoretical-textual) approach of phenomenology is required in order to provide appropriate micro sub-categories of surprise (and not a priori macro-categories) thanks to a reference to a situated first-person experience and description of a singular *hic et nunc* situation/moment where I am/was surprised. Thus a) the technics of explicitation microphenomenological interviews, which bring about first-person descriptions of a singular situation and which helps reading philosophical texts with an experiential look (EWR = Experiential Reading and Writing) and extracting concepts from texts as experienced ones.[66] And b) a generative comparison between a priori philosophical macro-categories and experiential micro-categories.

Here surprise has many advantages:

- Quoting Dennett, who writes it *here* remarkably: "Surprise is a wonderful dependent variable and should be used more often in experiments; it is easy to measure and is a telling betrayal of the subject's having expected something else (...) Surprise is only possible when it upsets belief."[67]

---

[65] On our current Emphline ANR Research project, the experiment uses four kinds of physiological measure: (1) cardiac frequency; (2) breath frequency; (3) skin conductivity; (4) cerebral pulsatility.

[66] N. Depraz, « Consciousness and First-Person Phenomenology: First steps towards an Experiential Phenomenological Writing and Reading (EWR), Sangetha Menon, Anindya Sinha & B. V. Sreekantan eds., *Interdisciplinary Perspectives on Consciousness and the Self*, Heidelberg, Springer, 2014 (from NIAS Conference in Bangalore, India, January 2012).

[67] D. Dennett, (2001) "Surprise, surprise," Commentary on O'Reagan and Noe. *Behav Brain Sci* 24(5), p. 982.

- Surprise as a startling, bodily cardiac reaction, which offers a strong experiential protentional-implicit anticipatory and aftermath-resonance continuity, the cardiac acceleration and deceleration dynamics being organically pre-conscious and not sub-consciousness like in the third-person braindynamics.
- Surprise is a pre-conscious first-person lived experience, which enters into narrow continuity with physiological data and resonates with quite an expressive bodily language accessible as re-lived through explicitation microphenomenological interviews.

2 Conclusion

The heart therefore appears not only as an integrative system but correlatively as a radiating system. It involves in its turn a multi-leveled dynamics based on coupling, valence, self-previousness, and a rainbowing of emotions. In this context, my contention is that surprise as a multivectorial process is the core-experience of a heart-centered model.

Part II
Surprise and Depression

The Temporal Dynamic of Emotional Emergence, Surprise and Depression

Thomas Desmidt

Abstract Surprise is often considered an intense but brief emotional reaction and many authors contend that surprise is a primary emotion, like joy or fear. However, surprise has some remarkable particularities that question the view of surprise as a primary emotion. Indeed, surprise can be either good or bad, i.e. of a positive or negative valence, whereas all the other primary emotions can only be of one valence, either good (e.g. joy) or bad (e.g. fear). Besides, surprise is rarely isolated from other emotions and rather often belongs to a global emotional dynamic, as surprise is generally followed by other emotions, happiness or deception for instance. Focusing on the neurophenomenological approach, the view of surprise as composed of a 3 phases-dynamic offers new insights in the science of emotions from experimental science, including psychophysiology and neuroscience, to phenomenology. We have shown some of these new insights in a recent article (Desmidt T, Lemoine M, Belzung C, Depraz N: Phenomenol Cogn Sci 14:557–578, 2014) and we want here to focus on their interests in the understanding of the psychopathology of depression and to show how the neurophenomenological methodology of co-generativity applied to the experience of surprise can lead to the emergence of new hypothesis that are consistent with both 1st and 3rd person-approaches in the understanding of the physiopathology and the psychopathology of depression, as a psychiatric disorder.

Keywords Temporal dynamic · Emotional emergence · Surprise · Depression · Psychopathology

T. Desmidt (✉)
UMR 1253, iBrain, Université de Tours, Inserm, Tours, France

CHU de Tours, Tours, France

N. Depraz, A. J. Steinbock (eds.), *Surprise: An Emotion?*, Contributions To Phenomenology 97, https://doi.org/10.1007/978-3-319-98657-9_3

1 Introduction

Surprise is often considered to be an intense but brief emotional reaction and many authors contend that surprise is a primary emotion, like joy or fear. However, surprise has some remarkable particularities that question the view of surprise as a primary emotion. Indeed, surprise can be either good or bad, i.e. of a positive or negative valence, whereas all the other primary emotions can only be of one valence, either good (e.g. joy) or bad (e.g. fear). Besides, surprise is rarely isolated from other emotions and rather often belongs to a global emotional dynamic, as surprise is generally followed by other emotions, happiness or deception for instance.

The phenomenological account of surprise also brings new perspectives. Focusing on the experience of surprise as a global experience, the phenomenological description of 'what is to be surprised' brings out the structures of the experience, i.e. how surprise emerges in the flow of the experiences and what are the preconscious features and conditions of possibility. This kind of approach led Varela & Depraz (Depraz 2008) to consider that surprise is not merely a brief reaction but that it always continues in a more or less long emotional phase that they called the aftermath. Moreover, they identified anticipation as a more or less conscious condition of the possibility of the experience of surprise (the anticipation is a necessary, although not sufficient, feature of surprise), as surprise is always a difference between a salient stimulus and some kind of anticipation. Through a phenomenological perspective, surprise is then composed of three temporal phases that are (1) anticipation, (2) crisis, and (3) aftermath, rather than being a brief reaction. Moreover, Depraz considered that surprise is essential to any other primary emotions and that it constitutes the threefold structure of the experience as well as of the dynamics of the emergence of emotion, consistent with Descartes' view of 'admiration' as the first emotion (Descartes 1989), that is, an experiential condition for every other emotions. Every emotion is then believed to emerge consistently with the temporal dynamic of surprise, and the different phases of this dynamic allow for the distinction of the different types of emotion.

Focusing on the neurophenomenological approach, the view of surprise as composed of a three phase-dynamic offers new insights in the science of emotions from experimental science, including psychophysiology and neuroscience, to phenomenology. We have shown some of these new insights in a recent article (Desmidt et al. 2014) and we want here to focus on their interests in the understanding of the psychopathology of depression, and to show how the neurophenomenological methodology of co-generativity applied to the experience of surprise can lead to the emergence of new hypothesis that are consistent with both 1st and 3rd person-approaches in the understanding of the physiopathology and the psychopathology of depression, as a psychiatric disorder.

2 The Three Phases of the Temporal Dynamic of Emotional Emergence

According to Husserl's conception of the living present, the temporal dynamic of experience is composed of retentions, impressions, and protentions that are acts of consciousness oriented toward the past, the present, and the future, respectively (Husserl 1991). These are pre-conscious processes that are dynamically acting (in a non-linear way but rather in parallel) to produce the lived time of the experience. Extending this conception to lived emotions and consistently to Varela's view of the specious present, Depraz (Depraz 2008) has proposed that during an emotional experience, this temporal dynamic is filled with more or less intense emotional fluctuations. Moreover, she proposed that the categorization of emotions relies on the temporal criterion that unfolds according to three phases and three scales of contexts. The three scales are (1) the pre-conscious, (2) the conscious, and (3) the intersubjective-historical level. Depending on the context scale, the temporal phases are called retention/impression/protention at the pre-conscious level, aftermath/crisis/anticipation at the conscious level and awaiting/event/commemoration at the intersubjective-historical level. Thus, anticipation corresponds to the expectation of a salient emotional stimulus to appear, the crisis occurs when the emotional stimulus is eventually perceived and the aftermath corresponds to the after-effects of the emotional stimulus that has disappeared and/or has been habituated. Interestingly, this two-vectors model allows for the identification of various emotions. For example, the anticipation, the crisis and the aftermath of a threat may be identified as the emotions of anxiety, fear and rumination, respectively. Alternatively, the anticipation, the crisis and the aftermath of a pleasant event may be identified as the emotions of 'looking forward', joy, and a 'euphoric state' (Table 1). Moreover, this temporal dynamic is not only operating in strong emotions but it shapes the emotional content of the lived time in general. Indeed, every salient event is like the more or less intense focal moment from which the three temporal phases of emotional emergence unfolds. We argue that in every moment, the experience is embedded in several parallel phases of more or less conscious anticipation, crisis, or aftermath. Even when it seems there are no emotional fluctuations, like in smooth coping when doing a repetitive action like hammering for instance, it happens in a

Table 1 The 3 phases of surprise and emotional experience with two examples of emotional experiences and the corresponding emotions distinguished as a function of the phases

	Anticipation	Crisis	Aftermath
	Expectation of an ES	Perception of an ES	Traces of the ES that has disappeared/ been habituated
ES=Threat	Anxiety	Fear	Ruminations
ES=Pleasant event	'looking forward' state	Joy	'euphoric' state

ES=emotional stimulus

particular context that is embedded in a temporal dynamic that includes not only a present component, but also a past and a future temporality which, for the latter, is often appearing as a certain purpose of the experience. We argue that the experience of hammering is embedded in the temporal dynamic of the present of hammering, but also the aftermath of an emotional event (the conception of the object to be crafted for example) and the anticipation of another emotional event, like when looking forward to finish the job either because it is boring and expecting its end, or because one would like to appreciate the final result. Moreover, one may be anxiously anticipating hitting the fingers! Therefore, and although it may indeed not include the phase of the crisis in smooth coping, it necessary includes some kind of aftermath and anticipation. Besides, when the phase of the crisis happens as a result of the end of the hammering process and the perception of the final version of the object (or when hitting a finger!), it reveals the whole dynamics of the 3 phases that may have remained preconscious until that time. In general, we argue that the temporal dynamic of experience unfolds according to three non-linear phases that are mostly appearing in the phase of the crisis that happens as a result of a salient/significant more or less unexpected event. Let's take an example of how the temporal dynamic of the three phases of emotional emergence is operating through an archetypical day. Let's begin, in the morning, by the crisis of the alarm clock that wakes one up. This crisis depends on how the night was (whether having a resting sleep or not for instance) and how it was anticipated (anxiously or looking forward to it for instance), as the wake up bell was anticipated in some way when it was set up in the previous evening. Then, comes the crisis of the shower, getting dressed, breakfast, etc… that may be pleasant or not, depending on the aftermath of the awakening and the anticipation of the day to come. Then comes the crisis on the way to the office that may also be pleasant or not, depending on the aftermath of the morning activities and the anticipation of the day of work to come. During work-time, crisis may occur, like a specific meeting involving a public speech for instance, and always taking place in a series of more or less pleasant aftermaths and anticipations that determine the intensity of the crisis. Finally, the crisis of the way back home happens again within aftermaths and anticipations, to finish with the crisis of the evening (that may be even more pleasant if the partner has prepared a surprised romantic dinner for instance!) and bedtime, eventually anticipating the crisis of the next morning! We argue that this overall dynamic of phases is the structure for the lived temporality.

3 The 1st (Phenomenology) and 3rd (Physiology) Accounts of the Three Phases

3.1 Anticipation

Let's take an example to illustrate our model. During a trek in the park ofYellowstone, one suddenly sees what looks like probable bear tracks. This stimulus is not per se salient, but it constitutes a cue stimulus that a salient stimulus can happen nearby and soon, what triggers an anxious anticipation of a threat, as a bear may somehow be dangerous.

The phenomenological account for this process of anticipation started with Husserl who described protention both as the presence of the future, or an openness toward the future anticipated under certain conditions in the form of 'awaiting', and as a tension (*Spannung*) toward the event to come (Husserl 1973). Then Heidegger proposed a similar position in *Being and Time* (Heidegger 1973) of an affect-laden future-oriented lived time. However, we believe that Heidegger's account can hardly be associated with third-person accounts, and we did not refer to his account for two reasons: (1) Heidegger's existential level of analysis is overly macro-embedded insofar as it hardly allows for the possibility of a fine-grained subjective description that can be mapped onto the micro-physiological and neuro-dynamics of time, which is the goal of our discussion of neurophenomenology; (2) Heidegger's account is primarily one-sided with a negative valence (e.g., anxiety and boredom). More recently, Varela argued for a generative role of protention and its emotional component (Varela 1999) and considered the experience of time, the 'being present,' the cultivation of the ability to anticipate the unexpected and to become aware of the strong emotional quality of such an 'unexpectation'.

The 3rd person accounts of anticipation have become more and more substantial in recent years. Indeed some experimental protocols allow for the investigation of anticipation, including the S1-S2 protocols. In these protocols, subjects successively perceive a cue stimulus (S1) that announces the imminent arrival of the S2 stimulus. Physiological changes have been investigated during the S1-S2 anticipation period. These physiological changes include (1) a negative wave in the electroencephalogram (EEG), known as the stimulus preceding negativity (SPN) (Brunia 1988); (2) an activation of specific brain regions detected using functional Magnetic Resonance Imaging (fMRI), including the insular cortex (Kotani et al. 2009); and (3) a specific triphasic pattern of Heart Rate (HR) changes (Simons et al. 1979) in which the perception of the S1 stimulus is immediately followed by a short deceleration (D1) that lasts approximately one or two seconds, followed by a short acceleration (A1) that also lasts approximately one or two seconds. Until the perception of the S2, the HR continuously decreases in a long and profound deceleration (D2). Whereas D1 and S1 are thought to be associated with the perception of the S1, D2 seems to be the most specific HR change of EA. Interestingly, experiments in which EA duration was manipulated have shown that the D2 HR deceleration lasted as long as the subject expected an emotional stimulus, whereas at the moment of the perception of the emotional stimulus, the HR stopped decelerating and immediately accelerated (Somsen et al. 2004). Moreover, the main determinant of the intensity of physiological changes seems to be the anticipated arousal level (Poli et al. 2007): the higher the arousal levels of the S2 (pictures of injuries or erotic couples), the larger the amplitude of the physiological changes.

3.2 Crisis

Let's continue with our example of the bear encounter. After some time of the anxious anticipation of the possible encounter with a bear, it eventually appears just in front of us and it is now the time for the crisis! Indeed, the current perception of an emotionally significant stimulus that was previously anticipated (either explicitly or implicitly) indicates the breakthrough from the emotional phase of anticipation to the emotional phase of the crisis, which generally corresponds to the climax of the emotional dynamic in terms of intensity. This phase begins with what is often called arousal, which is understood as an intense bodily reaction via the activation of the autonomous nervous system (ANS) (Belzung 2007). Moreover, one is not fully aware of one's emotion at the beginning of the crisis. Some duration exists between the beginning of the crisis and the awareness of the emotional dynamic. During this 'first' sub-phase of the crisis, one's awareness is disturbed by a bodily reaction. The subject is not fully aware of what happens, but he may feel some imprecise emotional fluctuations. We could say that the subject is imprecisely aware of his emotional state; that is, the subject feels that something is happening without being able to represent or report clearly what is occurring. He feels changes in his body, but he cannot feel that he is experiencing a clear and distinct type of emotion, such as sadness or joy. The 'first' sub-phase of the crisis is a non-lived time, a non-experience, in which there is no strict emotion, in the same way that Husserl (1973, §21) and Peirce (1958) contend that there is an interruption or a rupture, respectively, that creates a blank in the continuum of experience. Depraz (2008) calls this moment of the emotional dynamic a self-previousness period to indicate that one is affected before being fully aware of one's emotion. From a cognitive approach, we could say that the clear awareness of an emotion requires information from the immediate environment (through the perception of contextual stimuli), from the internal body (interoception), and from higher and more complex cognitive processes. This situation eventually leads to the emergence of a sense of the current situation, which allows for a *lived* emotion in the 'second' sub-phase of the crisis. The self-previousness period thus corresponds to the integration of information from the environment and the body, but not to higher cognitive processes; at the time, there is no clear and distinct representation of the situation, and no rational signification can arise. In contrast, when higher cognitive processes are integrated along with interoception and external perception, the experience of the emotions of fear, sadness, or joy can emerge in the 'second' sub-phase of the crisis.

The third-person accounts of the crisis have been extensive in the scientific literature and crisis has been associated with strong physiological reactivity. One study is of particular interest. Using the paradigm of emotional remembrance, some authors (Rainville et al. 2006) have shown patterns of cardiorespiratory activations that are specifically correlated with four different types of emotion (joy, sadness, fear, and disgust). This result is strikingly consistent with James's theory that there is a specific bodily reaction associated with every type of emotion. Other studies have investigated the neurobiological reactions involved in the perception of emotional stimuli. Interestingly, some areas, such as the right anterior insula and the adjacent inferior frontal operculum, are recruited during the second-order processing of one's internal bodily experience as well as during conscious emotional experience, which supports the theory that the perception of visceral changes is central to emotional experience (Craig 2002, 2009a; Singer et al. 2009).

3.3 Aftermath

Let's conclude our example of the bear encounter. Finally, the bear has quietly left causing no harm and it is now the time for the aftermath. The phenomenological account for aftermath starts with Husserl who reconsidered retention with a passive and dynamic meaning in applying his genetic phenomenology to perceptual experience (Husserl 2001, §35). Retention is thus understood in terms of 'resonance' as a consequence of the after-effects of an event. Resonance is neither cognitive nor corporal; it is structural. When resonance is filled with bodily reactions and cognitions, it becomes the emotional aftermath. Many types of emotions can occur in the aftermath phase, depending on the dynamic of emotional emergence. For example, Husserl (Husserl 1973) analyzed the aftermath of deception, which may follow the crisis of a shock and the anticipation of hope. Other examples include alleviation (if positive) or ruminations (if negative), which are the aftermath of a crisis of panic and an anticipation of anxiety. Similarly, an aftermath of euphoria can follow a crisis of wonder and an anticipation of curiosity. Moreover, the aftermath phase may be longer or shorter as a function of the intensity of the crisis, Finally, the aftermath phase, as part of a circular dynamic, may lead directly to the emergence of a new anticipation phase when the aftermath produces cues about an emotional event to come and allows the subject to reappraise new anticipation.

The third-person accounts of the aftermath have also been investigated to some extent and it has been associated with bodily reactions related to emotion and to the physiological process of recovery toward a basal state (Larsen and Christenfeld 2011). Moreover, some authors have proposed labeling the mental extension of stressors 'perseverative cognition,' a process that may be particularly involved in the development of stress-related diseases, including heart disease (Brosschot et al. 2006). Perseverative cognitions have been associated with increased HR and a decrease in heart rate variability (HRV) in some experimental conditions (after the expression or the inhibition of anger, it takes approximately

10 min for the heart system, as assessed by HRV, to recover and for HRV to return to its resting level (Dorr et al. 2007) as well as in ecological experimentation (for instance, on teachers for whom HR and HRV were continuously measured during their usual daily life (Pieper et al. 2010)). Interestingly, some authors have demonstrated that recovery from stressful anticipation (e.g., public speech) involves a similar cardiovascular profile as recovery from the stressful event itself (Waugh et al. 2010).

4 Surprise as the Structure of the Temporal Dynamic

Surprise may be understood in different ways. It is generally understood as a brief emotional reaction due to an unexpected stimulus. Surprise may also be understood as a rupture in the flow of time or as a startle reaction. In a broader acceptation, surprise can be seen as the basic component of every emotion (as it was conceptualized by Descartes for instance (Descartes 1989)), as joy or fear for example necessarily require surprise to emerge. We want to suggest that surprise may also be understood in a phenomenological way. We argue that surprise *is* the structure of the temporal dynamic of emotional emergence. Indeed, surprise happens when being in the presence of an unexpected emotional stimulus. Thus, surprise results from the more or less large difference of an anticipation and the occurrence of an emotional stimulus. According to this definition, then, there is no surprise without some kind of anticipation, which is thus a necessary condition for surprise. Moreover, although surprise can be understood as a brief reaction, the experience of surprise is always associated with a more or less long and intense aftermath. We argue then that surprise encompasses the three phases of the anticipation/crisis/aftermath and that, reciprocally, the dynamic of the three phases unfolds according to the temporality of surprise, understood as a lived experience which require the temporal structures of the three phases to occur. Surprise is not only the crisis of an emotion, nor the aftermath or the anticipation alone, but it encompasses the three phases in a global dynamic.

4.1 The Issue of Surprise Without Explicit Anticipation

While it is quite clear that surprise emerges from the difference between what I anticipate and what eventually happens, the issue of the emergence of a surprise without explicit anticipation is less obvious. We argue that without protention, no surprise is possible. When someone is completely caught by surprise, the emergence of the surprise is not due to the lack of protention but rather to the fact that what is expected during the protentional phase is completely different from what ultimately occurs, as conceptualized by authors such as Smith (1795). In fact, in a normal state, everyone continuously anticipates what should happen in similar situations, as conceptualized by Hume in his theory of induction and habit (Hume 1739) and by James on habit (James 1918). While walking on the street in a familiar context, I expect that this context will continue, and I expect no change in this context. It is because my anticipation is 'no change in the context' that a change in the context evokes a surprise. Following Husserl's view of protention, we argue that there is a very basic type of anticipation that relies on the lived body (*Leib*) as opposed to anticipation, which is a complex cognitive process. How can we characterize such a protention? The surprise elicited by the shout of 'Surprise!' is associated with the release of body modifications, such as a strong and rapid increase in heart and respiratory rates. These physical changes correspond to a change in interoception (the feelings from the body), from which information is delivered to the brain. I rapidly become aware that something has happened because my body has reacted and I feel surprise. However, some time is required before I become fully aware of what happened and my eventual experience of pleasure if I am facing a good friend. In that case, the surprise corresponds to the sudden change in interoception, and the protention corresponds to the recurrence of regular variations in my interoception. The hypothesis of protention is, thus, as follows: the recurrent variations of the physiological functions of the body constitute a familiar state in which the recurrence of identical variations is expected. An intense change in this recurrence constitutes a surprise.

Although, we do not aim to reduce surprise to merely a brief and intense reflex reaction with poor or no emotional properties, as we argue that surprise is a large process that includes three phases, it indeed includes some strong physiological reactions, namely the startle and the cardiac defense.

4.2 *The Startle Reflex and the Cardiac Defense*

The startle reflex and the cardiac defense are the two main reflexes associated with surprise as a lived experience in experimental protocols (Vrana et al. 1988). The experimental paradigm most commonly used to elicit these reflexes is exposure to an unexpected loud sound, generally white noise of 80 to 100 dB. A startle is a stereotyped brief and intense muscular reaction with a top-down gradient (initiated by contractions of the facial muscles, particularly the orbitofrontal muscles, whose contraction provokes a blink, which is the most consistent and persistent startle reflex, and continuing down to the lower limbs (Meinck 2006)). The startle reflex has been associated with a very small neural network and is not influenced by intentional control (Meinck 2006). Factors that are known to modulate the startle amplitude include attention, habituation, emotional context, anxiety, and mood disorders (Grillon and Baas 2003).

Cardiac defense is a stereotyped cardiac reflex characterized by a complex pattern of HR changes with accelerative and decelerative components (Vila et al. 2007). This reflex has a long tradition in psychophysiological research and is assumed to reflect the physiological reaction to a threat with the function of protection from potential injury. The pattern of HR changes elicited using a startle experiment is complex and it consists of four components across a duration of 80 s: (1) an initial short and intense acceleration that reaches its maximum in three seconds, followed by (2) a short and intense deceleration that reaches its maximum 15 s after the eliciting stimulus, followed by (3) a longer but less intense acceleration reaching its maximum at 35 s after the eliciting stimulus, and finally (4) a longer but less intense deceleration lasting until 80 s with a return to baseline HR (Vila et al. 2007). Known factors that modulate this HR pattern include attention, habituation, emotional context, anxiety disorders, gender, personality traits, and excessive worry. Some authors (Vila et al. 2007) have proposed that rather than a mere defense reaction, cardiac defense, which is composed of a complex pattern of HR changes with two accelerations and two decelerations, may reflect a combination of both cognitive and motivational reactions. This suggestion would mean that immediately after a surprise reaction, the first reflex component, as indexed by the first acceleration during the cardiac reflex, would be a defensive reaction with poor awareness of the situation. Only after 10–15 s would the physiological reactivity of the heart be sufficiently low to allow for attention to be focused on the context.

4.3 Synthesis of the Dynamic of Surprise as the Global Structure of Emotional Emergence

We contend that the complex physiological fluctuations observed in the pattern of HR changes and the mutual interaction of the startle, cardiac defense, and brain activations are the correlates of the emotional fluctuations of the dynamic of the three phases that lead to a distinct emotional experience, such as fear, sadness, or joy. The cardiac fluctuations reflect the phase of anticipation at both a physiological and an experiential level, from the perception of the cue stimulus to the 'awaiting' process, through the profound cardiac deceleration, and to the rapid acceleration that indicates the transition to the crisis. We contend that the startle and the beginning of the cardiac defense, which signal the beginning of the crisis, are the physiological correlates of the rupture, the blank in the continuum of experience described by Husserl and Peirce. In the aftermath, the resonance of surprise is seen in the fluctuations of cardiac physiology as an index of the mutual interaction of emotion and cognition.

Consistently with other authors (Thayer and Lane 2009), we contend that the cardiac physiology is an index of this integrative dynamic that includes the systems of the context, the body (and the heart and brain within the body), and the lived experience that dynamically interact according to the three phases to produce an emotional experience (Table 2).

Table 2 Configurations of the three systems (brain, context, body, and heart as the focal place of the body during emotion) involved in emotional emergence according to each of the three phases of the temporal dynamic (anticipation, aftermath, and crisis)

	Anticipation	Crisis	Aftermath
Context	Cue stimulus	Emotional stimulus	+/- Emotional stimulus
Body	Body schema associated with orienting	Startle and defensive action	Body schema associated with orienting and/or restoring
Heart	Pattern of HR changes associated with emotional anticipation	Early pattern of HR changes associated with cardiac defense	Late pattern of HR changes associated with cardiac defense, then cardiac recovery
Brain (with reference to Craig's model)	Global insular activation	Partial ('first' sub-phase) then global ('second' sub-phase) insular activation	Partial or global insular activation
Lived experience	Emotional fluctuations	Arousal and cognitive blur, blank event ('first' sub-phase), then *lived* emotion ('second' sub-phase)	Cognitive appraisal of the emotional context, then more or less perseverative cognitions

5 Toward an Integrative Model: The Temporal Dynamic of Emotional Emergence

5.1 A Heart-Centered Model as an Alternative for the Mind-Body Problem

In a recent article, Depraz (Depraz 2008) suggests that the brain-centered model should be displaced in favor of a heart-centered model in an attempt to move beyond the gap of the brain-mind problem. She contends that a heart-centered model could lead to an integrative model in which neurosciences, psychophysiology, and lived experience in a phenomenological sense are integrated, articulated, and co-generative in an effort to undo the remaining dichotomy between mind and brain, or the residual discontinuity between the phenomenal and the biological levels. Indeed, a heart-centered model could enable us to articulate the subpersonal, neural aspects of emotional mechanisms with the immanent, lived, expressive aspects of emotions at the subjective phenomenal qua pre-conscious level. In her article, Depraz extensively describes how the heart system can be understood as double-faced in the same way as the body system as a whole: it is both objective and subjective, physical and lived. The brain system, by contrast, remains one-sidedly physical and objective. For this reason, it is limited to the third-person approach, for which the mind or consciousness can only appear as 'correlative' (though irreducible) dimensions. In the heart system, which is more integrative because it is situated in the dynamic interface of the non-conscious (third-person)–conscious qua pre-conscious (first person), we find an experiential continuum between the body (working without the subject's knowledge) and the lived experience.

5.2 Heart Physiology and the Global Emotional Moment

In our view, an integrative model of emotional emergence would support discourses from the fields of phenomenology as well as physiology and neuroscience. Many models of emotion have been proposed in the neurosciences. However, we believe that the recent model proposed by Craig, a neuroanatomist, is of particular interest (Craig 2009a). The framework we have described in the previous chapters is strikingly consistent with his model, particularly with two of our major arguments: (1) the heart acts before one's awareness and, as such, sets the stage for emotional experience; and (2) the heart is a key feature that contributes to the temporalization of emotional experience.

In Craig's view, an emotional experience (and what he calls a global emotional moment, GEM) is produced by the sequential integration in the insular cortex of five types of information according to a spatial gradient. Interestingly, this spatial gradient is consistent with the temporal dynamic we have described in the previous sections. Indeed, the first integration (spatially in Craig's view and temporally in our view) concerns primary interoception (especially via the vagal nerve, whose fibers are 80% afferent (Jänig 1996)), among which the heart and the lung are believed to be the most important systems according to the homeostatic theory (Jänig 2006). This is followed by motor information (proprioception), environmental information (perception of the context), hedonic information (valence), and, finally, complex cognitive information, such as social information.

Moreover, Craig suggests that the layout of the GEM accounts for the time dilatation or restriction that occurs during an emotional experience (Craig 2009b). A negative emotional experience is associated with the synthesis of a greater number of GEMs (resulting in the feeling of accelerated time), and a positive emotional experience would be associated with the synthesis of a lower number of GEMs (resulting in the feeling of time passing more slowly). Although the precise mechanism of this processing remains to be investigated, Craig suggests that impulses from the cardiac system could be a key feature of the temporal indexation of the GEM. Ultimately, this finding is consistent with the idea that the core of the temporal dynamic in emotion may be the physiology of the heart in its interaction with the brain.

6 Surprise and Depression

In the last section, we would like to show how the model of the temporal dynamic of emotional emergence can be heuristic for the understanding of emotional reactivity in depression. Depression is a psychopathological disorder that includes emotional, cognitive and behavioral symptoms. The two main symptoms for depression are intense and pervasive sadness and anhedonia (hyporeactivity to pleasure). Thus, depression can be understood as an emotional disorder but the true nature of emotional impairment in depression is not yet fully known. A recent meta-analysis (Bylsma et al. 2008) suggests that depression is associated with a global emotional reactivity, i.e. both for positive and negative stimulus. Other authors consider that depression is associated with a decreased reactivity to positive stimulus and an increased reactivity to negative stimulus. The most probable explanation for inconsistent results regarding reactivity to negative stimulus in depression is the high heterogeneity of the depressive syndromes and the frequent comorbidity with anxiety that is associated with hyper-reactivity to negative stimulus (Bylsma et al. 2008). Thus, depression may be characterized by either a global or partial emotional hyporeactivity.

In our model of emotional emergence, an emotional hyporeactivity would refer to a decreased reactivity to surprise as a global dynamic and to the three phases of anticipation/crisis/aftermath. Some recent studies (Sherdell et al. 2012) suggest that anhedonia may not be characterized by an impairment of pleasure consummation but rather by an anticipation of pleasure, that is depressive patients may primarily lack the possibility to anticipate pleasure outcomes whereas they may eventually feel pleasure when they are in direct contact with the pleasant stimulus. For example, a depressive subject may not be able to anticipate the pleasure to savor a good meal but when the food is eventually in his mouth he may say that the food is good. In reference to our model, hyporeactivity in depression may be primarily understood as an impaired emotional anticipation which is a necessary, but not sufficient, condition for surprise and emotional emergence. Moreover, as the heart physiology is an index of emotional emergence in our model, hyporeactivity in depression may be characterized by an impaired cardiac physiology especially during the anticipation phase. Furthermore, cardiac and cerebrovascular disorders are now well known to be associated with depression (Camus et al. 2004; Nemeroff and Goldschmidt-Clermont 2012). For example, the first cause of death in depressive patients is cardiovascular diseases. Besides, depression is recognized as a strong and independent cardiovascular risk factor (Baxter et al. 2011; Charlson et al. 2011), especially for myocardial infarction, similarly to (but independently from) diabetes, smoking status, hypercholesterolemia. In addition, the depression/cardiovascular diseases association is bidirectional (Camus et al. 2004) which means that depression is a cardiovascular risk factor, but that cardiovascular diseases also constitute a risk for depression as depression is highly prevalent in patients with myocardial infarction or cerebrovascular diseases. However, the physiopathology of this association remains unknown (Nemeroff and Goldschmidt-Clermont 2012), although several processes have been involved (Stapelberg et al. 2011). Especially, low heart rate variability (HRV) and an endothelial dysfunction have been associated with depression. HRV is the measure of the HR changes as a function of time and can be assessed with many indexes that are believed to represent the ANS activity. As a general rule, a low HRV (low amplitude of HR variations) reflects bad health and poor cardiovascular outcomes (Thayer et al. 2010). Endothelial function (the endothelium is the arteries layer in contact with blood) is believed to reflect the reactivity of arteries which require adaptation to the blood flow changes. An endothelial dysfunction reflects poor arteries reactivity and poor cardiovascular outcomes. Both low HRV and endothelial dysfunction have been associated with depression (Isingrini et al. 2009) although results are inconsistent and again the clinical heterogeneity of depression is a probable explanation. As cardiovascular physiology is a main index of emotional emergence in our model and since hyporeactivity in depression may be primarily driven by an impaired emotional anticipation, we argue that cardiovascular physiology is primarily impaired in emotional anticipation in depressive patients.

However, this statement remains a hypothesis that requires experimental verifications and we are currently implementing an experimental protocol to confirm (or not) this hypothesis. In this study (the Emphiline project (Depraz, N. (in preparation)), we aim to show that emotional processes are at the core of the link between depression and cardiovascular diseases and that surprise and anticipation impairment in depression can be characterized by a hyporeactivity of cardiac and cerebrovascular physiology. We designed a protocol in order to assess the cardiac and cerebrovascular physiology during the three phases of emotion with a S1-S2 paradigm, involving emotional pictures, in depressive patients as well as controls and patients remitted from depression (in order to show whether emotional hyporeactivity is a trait or state marker of depression). During the three phases, the following will be recorded: the HR and respiratory rate changes, the ANS reactivity, the startle amplitude, and the brain pulsatility (which we have shown to be dramatically decreased in depressive patients at rest (Desmidt et al. 2011)) as an index of brain endothelial function. If we could make an association with cardio and cerebrovascular hypo-reactivity (decreased amplitude of HR and brain pulsatility changes) and depression during anticipation, we could argue that HR changes and brain pulsatility are indexes of the temporal dynamic of emotional emergence. Moreover, patients will be interviewed according to the explicitative interview methods to collect rigorous first-person data and to do a dynamical association with third-person data in order to characterize the phenomenological nature of emotional reactivity in depression.

7 Conclusion

The neurophenomenological project aimed at mixing phenomenological and neuroscientific accounts in a dynamical way of co-generativity to produce new insights in cognitive sciences. According to this project, we have summarized the model of the temporal dynamic of emotional emergence that involves the mutual interaction of the pre-conscious (the physiology of the heart and the brain) and the conscious (lived experience) levels according to a non-linear dynamic of three temporal phases. We have shown how these phases follow the dynamic of surprise because they consist of the emotional shock we call the crisis, which is only possible through anticipation and which ultimately continues to the aftermath. Moreover, we have shown how this neurophenomenological framework could lead to new hypothesis on the pathophysiology of depression for an integrative understanding of it and for the characterization of the emotional emergence of depression both at the first- and third-person levels. Along with experimental evidences, this would ultimately lead to the rigorous identification of both first-person (especially thanks to the explicitative interviews methodology) and third-person markers for depression, and to the development of therapy related to these markers, probably including the rehabilitation of surprise in depression both in the lived experience and in the cardiac and brain physiology.

References

Baxter, A.J., F.J. Charlson, A.J. Somerville, and H.A. Whiteford. 2011. Mental disorders as risk factors: Assessing the evidence for the global burden of disease study. *BMC Medicine 9* (1): 134. https://doi.org/10.1186/1741-7015-9-134.

Belzung, C. 2007. *Biologie des émotions*. De Boeck Supérieur.

Brosschot, J.F., W. Gerin, and J.F. Thayer. 2006. The perseverative cognition hypothesis: A review of worry, prolonged stress-related physiological activation, and health. *Journal of Psychosomatic Research 60* (2): 113–124.

Brunia, C.H. 1988. Movement and stimulus preceding negativity. *Biological Psychology 26* (1-3): 165–178. Accessed 24 September 2011.

Bylsma, L.M., B.H. Morris, and J. Rottenberg. 2008. A meta-analysis of emotional reactivity in major depressive disorder. *Clinical Psychology Review 28* (4): 676–691. https://doi.org/10.1016/j.cpr.2007.10.001.

Camus, V., H. Kraehenbühl, M. Preisig, C.J. Büla, and G. Waeber. 2004. Geriatric depression and vascular diseases: What are the links? *Journal of Affective Disorders 81* (1): 1–16. https://doi.org/10.1016/j.jad.2003.08.003.

Charlson, F.J., N.J.C. Stapelberg, A.J. Baxter, and H.A. Whiteford. 2011. Should global burden of disease estimates include depression as a risk factor for coronary heart disease? *BMC Medicine 9*: 47. https://doi.org/10.1186/1741-7015-9-47.

Craig, A.D. 2002. How do you feel? Interoception: The sense of the physiological condition of the body. *Nature Reviews. Neuroscience 3* (8): 655–666.

———. 2009a. How do you feel--now? The anterior insula and human awareness. *Nature reviews. Neuroscience 10* (1): 59–70.
———. 2009b. Emotional moments across time: A possible neural basis for time perception in the anterior insula. *Philosophical Transactions of the Royal Society of London Series B, Biological Sciences 364* (1525): 1933–1942.
Depraz, N. 2008. The rainbow of emotions: At the crossroads of neurobiology and phenomenology. *Continental Philosophy Review 41*: 237–259. Accessed 24 September 2011.
———. (in preparation) - Pays Germanique - UMR 8547 "les archives Husserl de Paris", ENS-CNRS, Paris, France. (n.d.). http://www.umr8547.ens.fr/spip.php?rubrique50
Descartes, R. 1989. *The Passions of the Soul: Les Passions De l'Âme*. (S. Voss, Trans.). Indianapolis: Hackett Publishing Company, Inc.
Desmidt, T., M.E. Hachemi, J.-P. Remenieras, P. Lecomte, N. Ferreira-Maldent, F. Patat, and V. Camus. 2011. Ultrasound Brain Tissue Pulsatility is decreased in middle aged and elderly type 2 diabetic patients with depression. *Psychiatry Research 193* (1): 63–64. https://doi.org/10.1016/j.pscychresns.2011.01.002.
Desmidt, T., M. Lemoine, C. Belzung, and N. Depraz. 2014. The temporal dynamic of emotional emergence. *Phenomenology and the Cognitive Sciences*: 1–22. https://doi.org/10.1007/s11097-014-9377-8.
Dorr, N., J.F. Brosschot, J.J. Sollers 3rd, and J.F. Thayer. 2007. Damned if you do, damned if you don't: The differential effect of expression and inhibition of anger on cardiovascular recovery in black and white males. *International Journal of Psychophysiology 66* (2): 125–134.
Grillon, C., and J. Baas. 2003. A review of the modulation of the startle reflex by affective states and its application in psychiatry. *Clinical Neurophysiology 114* (9): 1557–1579.
Heidegger, M. 1973. *Being and Time*. New York: Basil Blackwell.
Hume, D. 1739. *A Treatise of Human Nature*. Digireads.com Publishing.
Husserl, E. 1973. *Experience and Judgment*. Evanston: Northwestern University Press.
———. 1991. *Collected Works: On the phenomenology of the consciousness of internal time, 1893–1917*. Dordrecht: Kluwer Academic Pub.
———. 2001. *Analyses concerning passive and active synthesis: Lectures on transcendental logic*. Dordrecht: Springer.
Isingrini, E., T. Desmidt, C. Belzung, and V. Camus. 2009. Endothelial dysfunction. A potential therapeutic target for geriatric depression and brain amyloid deposition in Alzheimer's disease? *Current Opinion in Investigational Drugs (London, England: 2000) 10* (1): 46–55. Accessed 10 Sept 2011.
James, W. 1918. *The principles of psychology*. H. Holt.
Jänig, W. 1996. Neurobiology of visceral afferent neurons: Neuroanatomy, functions, organ regulations and sensations. *Biological Psychology 42* (1-2): 29–51. Accessed 5 Sept 2012.
———. 2006. *Integrative action of the autonomic nervous system*. Cambridge: Cambridge University Press.
Kotani, Y., Y. Ohgami, Y. Kuramoto, T. Tsukamoto, Y. Inoue, and Y. Aihara. 2009. The role of the right anterior insular cortex in the right hemisphere preponderance of stimulus-preceding negativity (SPN): An fMRI study. *Neuroscience Letters 450* (2): 75–79. Accessed 14 Aug 2012.
Larsen, B.A., and N.J.S. Christenfeld. 2011. Cognitive distancing, cognitive restructuring, and cardiovascular recovery from stress. *Biological Psychology 86* (2): 143–148.
Meinck, H.-M. 2006. Startle and its disorders. *Clinical Neurophysiology 36* (5-6): 357–364.
Nemeroff, C.B., and P.J. Goldschmidt-Clermont. 2012. Heartache and heartbreak-the link between depression and cardiovascular disease. *Nature Reviews. Cardiology 9* (9): 526–539. https://doi.org/10.1038/nrcardio.2012.91.
Peirce, C.S. 1958. *Collected papers of Charles Sanders Peirce*. Cambridge, MA: Harvard University Press.
Pieper, S., J.F. Brosschot, R. van der Leeden, and J.F. Thayer. 2010. Prolonged cardiac effects of momentary assessed stressful events and worry episodes. *Psychosomatic Medicine 72* (6): 570–577.

Poli, S., M. Sarlo, M. Bortoletto, G. Buodo, and D. Palomba. 2007. Stimulus-Preceding Negativity and heart rate changes in anticipation of affective pictures. *International Journal of Psychophysiology 65* (1): 32–39. Accessed 24 Sept 2011.

Rainville, P., A. Bechara, N. Naqvi, and A.R. Damasio. 2006. Basic emotions are associated with distinct patterns of cardiorespiratory activity. *International Journal of Psychophysiology 61* (1): 5–18.

Sherdell, L., C.E. Waugh, and I.H. Gotlib. 2012. Anticipatory pleasure predicts motivation for reward in major depression. *Journal of Abnormal Psychology 121* (1): 51–60. https://doi.org/10.1037/a0024945.

Simons, R.F., A. Ohman, and P.J. Lang. 1979. Anticipation and response set: Cortical, cardiac, and electrodermal correlates. *Psychophysiology 16* (3): 222–233. Accessed 24 Sept 2011.

Singer, T., H.D. Critchley, and K. Preuschoff. 2009. A common role of insula in feelings, empathy and uncertainty. *Trends in Cognitive Sciences 13* (8): 334–340.

Smith, A. 1795. SECTION I: Of the Effect of Unexpectedness, or of Surprise. In *Essays on Philosophical Subjects* (W. P. D. Wightman and J. C. Bryce, vol. III of the Glasgow Edition of the Works and Correspondence of Adam Smith (Indianapolis: Liberty Fund, 1982).).

Somsen, R. J. M., Jennings, J. R., & Van der Molen, M. W. (2004). The cardiac cycle time effect revisited: Temporal dynamics of the central-vagal modulation of heart rate in human reaction time tasks. *Psychophysiology*, *41*(6), 941–953.

Stapelberg, N.J.C., D.L. Neumann, D.H.K. Shum, H. McConnell, and I. Hamilton-Craig. 2011. A topographical map of the causal network of mechanisms underlying the relationship between major depressive disorder and coronary heart disease. *The Australian and New Zealand Journal of Psychiatry 45* (5): 351–369. https://doi.org/10.3109/00048674.2011.570427.

Thayer, J.F., and R.D. Lane. 2009. Claude Bernard and the heart-brain connection: Further elaboration of a model of neurovisceral integration. *Neuroscience and Biobehavioral Reviews 33* (2): 81–88. Accessed 24 September 2011.

Thayer, J.F., S.S. Yamamoto, and J.F. Brosschot. 2010. The relationship of autonomic imbalance, heart rate variability and cardiovascular disease risk factors. *International Journal of Cardiology 141* (2): 122–131. Accessed 5 Sept 2012.

Varela, F. 1999. The specious present: A neurophenomenology of time consciousness. In *Naturalizing phenomenology: Issues in contemporary phenomenology and cognitive science.* Stanford University Press.

Vila, J., P. Guerra, M.A. Muñoz, C. Vico, M.I. Viedma-del Jesús, L.C. Delgado, et al. 2007. Cardiac defense: From attention to action. *International Journal of Psychophysiology 66* (3): 169–182.

Vrana, S.R., E.L. Spence, and P.J. Lang. 1988. The startle probe response: A new measure of emotion. *Journal of Abnormal Psychology 97* (4): 487–491.

Waugh, C.E., S. Panage, W.B. Mendes, and I.H. Gotlib. 2010. Cardiovascular and affective recovery from anticipatory threat. *Biological Psychology 84* (2): 169–175.

Animal and Human Models of Startle, Emotion, and Depression

Bruno Brizard

Abstract Emotion is an essential feature of both animal and human behavior. Film industry has well understood the impact of emotions on human experiences and has applied strategies that induce the viewer to feel strong emotions. The startle effect is the main tool used as "the ultimate form of sensationalism, artlessness, and mindlessness…[it] reaffirm[s] the animalistic, the atavistic, and the irrational expressions".

Here we would like to show how startle is linked to both human and animal emotions, and how experimental studies can inform us on this relation. We will also give an account on how the animal models are used to characterize the neurobiological features of depression and how startle is impaired depression as shown in experimental studies.

Keywords Animal · Human models · Startle · Emotion · Depression

1 Introduction

Emotion is an essential feature of both animal and human behavior. The film industry has well understood the impact of emotions on human experiences and has applied strategies that induce the viewer to feel strong emotions. The startle effect is the main tool used as "the ultimate form of sensationalism, artlessness, and mindlessness…[it] reaffirm[s] the animalistic, the atavistic, and the irrational expressions."[1]

Here we would like to show how startle is linked to both human and animal emotions, and how experimental studies can inform us about this relation. We will also give an account of how the animal models are used to characterize the neurobiological

[1] Robert Baird, "The Startle Effect," in *Film Quarterly* 53, no. 3 (2000): 1–12.

B. Brizard (✉)
Inserm University of Tours, Tours, France
e-mail: bruno.brizard@univ-tours.fr

N. Depraz, A. J. Steinbock (eds.), *Surprise: An Emotion?*, Contributions To Phenomenology 97, https://doi.org/10.1007/978-3-319-98657-9_4

features of depression and how startle, as demonstrated in experimental studies, is impaired depression.

2 Emotion

2.1 Current Leading Models of Emotions

The attempt to define emotions is currently a challenge. However, some components of emotions have been recurrently suggested. Among the three components classically described, the first is called the situational component. It is usually considered the beginning of the process of emotion and is related to the contextual aspects that trigger emotions. The second component is called the cognitive component, and it relates to the interpretation of these events and situations. The third component is the physiological component that refers to the activation of regulation systems.[2]

Further developing this view of emotion as a three-components process, Gross et al. depict the spectrum of emotions in three phases that are part of a timeline.[3] The timeline evolves from basic emotions to emotional evaluation, thus inducing a psychological structure. Characterizing this structure then enables us to deduce emotional self-regulation. The final stage, then, is the confrontation between the social and the psychological structures, implying this emotional self-regulation. In other words, Gross establishes the fact that basic emotions are the starting point of this process that is constituted by a brief but efficient internal exchange (cerebral and peripheral) that produces the emotional experience. This will then lead to the elaboration of a strategy of reactions that will ultimately be integrated into the social context. Moreover, to be ready to cope with its environnement the whole process requires anticipation, and the brain needs to anticipate in order to adjust *to* the emotional reaction itself.

Furthermore, emotions are generally distinguished into basic and secondary emotions, the first referring to situations requiring instinctual reactivity alone (for instance, survival or reproduction[4]) and the second to a more elaborated process.[5] Plutchik is famous for his characterization of basic and secondary emotions and suggests a multidimensional model.[6] It is based on eight basic emotions evolving

---

[2] J.J. Gross and L. Feldman Barret, "Emotion Generation and Emotion Regulation: One or Two Depends on Your Point of View," *Emotion Review* 3, no. 1 (2011): 8–16. doi: https://doi.org/10.1177/175403910380974

[3] Ibid.

[4] Michiko Sakaki, Kazuhisa Niki, and Mara Mather. "Beyond Arousal and Valence: The Importance of the Biological versus Social Relevance of Emotional Stimuli." *Cognitive, Affective, & Behavioral Neuroscience* 12, no. 1 (March 1, 2012): 115–39. doi:https://doi.org/10.3758/s13415-011-0062-x

[5] Gross and Barrett, "Emotion Generation and Emotion Regulation."

[6] Robert Plutchik and Henry Kellerman, *Theories of Emotion* (Cambridge, MA: Academic Press, 2013).

from more complex emotions by association. His model is composed of three components: the dimensional component (intensity, similitude and polarity), the persistence component (acute or chronic), and the purity component. Basic emotions are believed to emerge from the dynamic of these components. According to Plutchik, these emotions are fear, surprise, sadness, disgust, anger, anticipation, joy and trust. Ekman has also suggested a model of emotions and has presented emotions in a more expressive way, making a distinction between different emotions by studying the facial muscle activity.[7] He actually extricates seven emotions, which he calls them universal emotions, that follow identical patterns across different cultures. These are fear, anger, sadness, happiness, disgust, surprise, and contempt.

The number and type of the basic emotions are a matter of debate, but surprise is generally included in the basic emotions. However, surprise appears to be very difficult to characterize. The emotional state of mind linked to surprise may be presented as an initial state of mind that is suddenly disrupted by something unexpected or predicted. Then comes another emotion depending of the valence (neutral/moderate, pleasant/unpleasant, positive/negative) and the intensity.

2.2 *Animal Models of Emotions*

Since ancient times, animal models for research have led us to a better understanding of human functionning, especially of our brain.[8] Animal models allow us to investigate neurobiological processes that may remain uneasy to investigate in humans. Animal models have proven to be particularly useful for understanding psychiatric disorders, including neurobiological pathways and pharmacological therapies. However, animal models can only be effective if they follow the following requirements: homological validity, pathogenic validity, mechanistic validity, face validity, and predictive validity.[9] Moreover, the most adequate species must be used in order to meet the following skills:

- Logistic systems (genes, nervous system, monoamines, neuroendocrine systems)
- Cognitive aptitudes associated with the emotions[10]

---

[7] Paul Ekman and Erika L. Rosenberg, *What the Face Reveals: Basic and Applied Studies of Spontaneous Expression Using the Facial Action Coding System (FACS)* (Oxford: Oxford University Press, 2005).

[8] Aristotle, *History of Animals* (Cambridge, MA: Harvard University Press, 1965).

[9] Catherine Belzung and Maël Lemoine, "Criteria of Validity for Animal Models of Psychiatric Disorders: Focus on Anxiety Disorders and Depression," in *Biology of Mood & Anxiety Disorders* 1, no. 1 (2011): 9. doi:https://doi.org/10.1186/2045-5380-1-9

[10] Catherine Belzung and Pierre Philippot, "Anxiety from a Phylogenetic Perspective: Is There a Qualitative Difference between Human and Animal Anxiety?" *Neural Plasticity 2007* (2007): 59676. doi:https://doi.org/10.1155/2007/59676

- Logistic systems (genes, nervous system, monoamines, system neuroendocrinien)
- Cognitive aptitudes associated with the emotions (cognitive bias)
- Behavior (emotional expression, physiological component

Rodents and monkeys have been shown to be adequate species for the study of emotional processes. For example, Berridge et al. have found similitaries in facial reactions between the human baby, the orangutan, and the rat in a context confronting like and dislike reaction tastes.[11] Berridge has shown that this type of behaviour is qualified as interspecies and is innate. Moreover, rodents appear quite easily manipulated and have proved to have a physiology related to emotions very similar to humans, including gene expressions.[12] Rodents have also proved to be very interesting for testing the effects of drugs, including the effect of antidepressants to reverse a depressive like state.[13]

3 Startle in Experimental Studies[14]

The startle reaction is believed to be related to surprise as a basic reaction. In the classical startle procedure, the subject receives a strong sound stimulation. We can then study the neurobiological processes related to this stimulation. Since it was first proposed in 1929,[15] the acoustic startle protocol has benefited several technological advances that have led to a more accurate definition of this phenomenon. The observation[16] of the motor activity related to muscular production is picked up by Ekman in his analysis of emotional muscular activation.[17] Ekman was able to prove that, contrary to other emotions, the phenomenon provoked by the startle procedure is impossible to simulate and that its reactivity has a considerably shorter span. It does not, however, appear to be a simple reflex, since this phenomenon can

---

[11] Kent C. Berridge and Piotr Winkielman, "What Is an Unconscious Emotion? (The Case for Unconscious 'Liking')," *Cognition & Emotion* 17, no. 2 (March 2003): 181.

[12] Christopher JL. Murray and Alan D Lopez, "Alternative Projections of Mortality and Disability by Cause 1990–2020: Global Burden of Disease Study," *The Lancet* 349, no. 9064 (May 24, 1997): 1498–1504. doi:https://doi.org/10.1016/S0140-6736(96)07492-2

[13] Belzung and Philippot, "Anxiety from a Phylogenetic Perspective."

[14] Dreissen, Yasmine E.M., Mirte J. Bakker, Johannes H.T.M. Koelman, and Marina A.J. Tijssen. "Exaggerated Startle Reactions." *Clinical Neurophysiology* 123, no. 1 (January 2012): 34–44. doi:https://doi.org/10.1016/j.clinph.2011.09.022

[15] H. Strauss H, "Das zusammenschrecken: Experimentell-Kinematographische Studie zur Physiologie und Pathophysiologie der Reaktivbewegungen," in *J Psychol Neurol* 39 (1929): 111–231.

[16] C. Landis and Hunt W, *The Startle Pattern* (New York: Farrar & Rinehart, Inc., 1939).

[17] Ekman and Rosenberg, *What the Face Reveals.*

be adjusted to different levels of intensity[18] according to the emotional context.[19] However, Ekman's analysis was based on camera recordings and did not allow for a precise measure of the muscular activity, showing its limitations, for instance, as the subject closed its eyes when a gun shot produced a stimulating effect.[20]

Another technique proposed by Davis in 1948[21] used the electromyographic recording. But it is really during the 1970s, with researchers like Gogan,[22] Rossignol,[23] and Fox,[24] that this method started to spread in the scientific community; it was finally standardized by combining a calibrated, hearing stimulation (between 90 and 110 db). Brown et al.[25] further developed this standardization by using an auditory tone burst with a frequency of 1000 hz, 50 ms duration, and 124 db, during which EMG activity of the oricularis oculi muscle is recorded, as this muscle has been shown to be the fastest to respond to the stimulus (20–50 ms), and it was seen as an early defensive posture.[26] Thanks to this approach, we are now able to understand the startle reaction as a mimicking of surprise. It shows that it is the first event of an emotional process and it can be now observed as the primary emotional expression.

[18] M.M. Bradley and P. J. Lang, "Affective Reactions to Acoustic Stimuli," in *Psychophysiology* 37, no. 2 (March 2000): 204–15.

[19] Gabriel S. Dichter, Andrew J. Tomarken, and Brian R. Baucom, "Startle Modulation Before, during and after Exposure to Emotional Stimuli," in *International Journal of Psychophysiology* 43, no. 2 (2002): 191–96.

[20] Ekman and Rosenberg, *What the Face Reveals*.

[21] R. Davis, "Motor effects of strong auditory stimuli," in *J Experimental Psychology* 38, no. 3 (1948):257–75. EM del Giudice, G. Coppola, G. Bellini, G. Cirillo, G. Scuccimarra, and A. Pascotto A, "A mutation (V260 M) in the middle of the M2 pore-lining domain of the glycine receptor causes hereditary hyperekplexia," in *Eur J Hum Genet* 9, no. 11 (2001): 873–6.

[22] P. Gogan, "The startle and orienting reactions in man. A study of their characteristics and habituation," in *Brain Res* 18 (1970): 117–35.

[23] S. Rossignol, "Startle responses recorded in the leg of man," in *Electroencephalography and Clinical Neurophysiology* 39, no. 4 (1975): 389–97.

[24] JE Fox, "Excitatory and inhibitory components of the eyeblink responses to startle evoking stimuli, studied in the human subject," in *Electroencephalography and Clinical Neurophysiology* 44, no. 40 (1978): 490–501.

[25] P. Brown, JC Rothwell, PD Thompson, TC Britton, BL Day, and CD Marsden, "New observations on the normal auditory startle reflex in man," in *Brain* 114 (1991): 1891–902.

[26] Ye. N. Sokolov, *Perception and the Conditioned Reflex* (Elmsford, NY: Pergamon Press, 1963). Also, P. Brown, BL Day, JC Rothwell, PD Thompson, and CD Marsden, "The effect of posture on the normal and pathological auditory startle reflex," in *J Neurol Neurosurg Psychiatry* 54 (1991): 892–7.

4 The Startle Studies in Depression

4.1 Depression and Human Studies of the Startle

As an affective disorder, depression can be understood as an impaired emotional reactivity. This disorder can be tested with the startle protocol. Kaviani et al. have noticed a strong dependency between the valence of the environment and the degree of intensity of the stimulus caused by startle. A pleasant environnement is associated with a milder response compared to a neutral one. On the contrary, an unpleasant environment will lead to stronger responses. When applied to depression, however, the context sensitivity does not allow us to distinguish patients from healthy subjects.[27] However, and since depression is characterized by heterogeneous behavioral and cognitive features, the severity of the depressive episode distinguishes low and high reactivity to startle, in which lower reactivity is associated with a higher severity of depression. Moreover, a higher level of anhedonia is associated with lower startle reactivity, whereas a high level of anxiety is associated with strong startle reactivity. Furthermore, some studies have found a temporality in the startle reactivity depending on the valence: the starle reactivity appears to be stronger in the anticipation of negative stimulus (though this has not been applied to depression yet).[28]

4.2 Starle and Animal Model of Depression

Startle has been extensively studied in animals (especially in rodents) since the late 30's.[29] The neurobiology of the rodent's startle has been well characterized and include a three-step neural pathway (cochlear root nucleus, pontine reticular nucleus, motor neurones) embedded in several connections. The startle paradigm is believed to be a reliable tool to assess emotional reactivity, and it has been shown to be modulated by affective states.[30] These studies found a strong link between the basic neural startle pathways and neural structures involved in emotional processing, including the amygdala, the prefrontal cortex, and the hippocampus.

---

[27] H. Kaviani, J.A. Gray, S.A. Checkley, P.W. Raven, G.D. Wilson, and V. Kumari, "Affective Modulation of the Startle Response in Depression: Influence of the Severity of Depression, Anhedonia, and Anxiety," *Journal of Affective Disorders* 83, no. 1 (November 15, 2004): 21–31. doi:https://doi.org/10.1016/j.jad.2004.04.007

[28] Dichter et al., "Startle Modulation."

[29] Landis and Hunt, *The Startle Pattern*. Also, Koch, M. "The Neurobiology of Startle." *Progress in Neurobiology* 59, no. 2 (October 1999): 107–28. doi:https://doi.org/10.1016/S0301-0082(98)00098-7

[30] Christian Grillon and Johanna Baas, "A Review of the Modulation of the Startle Reflex by Affective States and Its Application in Psychiatry," in *Clinical Neurophysiology* 114, no. 9 (September 2003): 1557–79. doi:https://doi.org/10.1016/S1388-2457(03)00202-5

The animal model of depression has also been shown to modulate the startle reflex. The most commonly used animal model of depression is based on the methodology of the Unpredictible Chronic Mild Stress (UCMS) technique that has been demonstrated to induce a state related to depression.[31] During UCMS, animals are exposed to repeated minor-intensity stressors, including environnemental stress (cage change, damp sawdust, cage tilting, predator sounds, etc.) at unpredictable times over several weeks, in order to elicit a number of behavioral alterations related to anhedonia and apathy. UCMS has been shown to have good validity based on the criteria relevant for human depression (see above) when used on rodents (Surget et al.). In depressive rodents, startle amplitude is lower,[32] consistent with the view of a decreased emotional reactivity in depression.

To date, most of the neurobiological studies on startle have used lesional and pharmacological blockage. The very recent optogenetic methodology brings new potentialities to investigate selective neurobiological mechanisms. Developed in 2003, the optogenetic technique combines "tissue and cell-specific expression of light sensitive microbial proteins called opsins along with advanced optical method to reach and control the activity of specific populations of cells in-vitro and in living animals with high temporal precision and quick reversibility."[33] In 2013, Ramirez et al. were able to manipulate the memories of a rodent in a fearful environment using optogenetics.[34] Ramirez demonstrated that it is possible to create an internal representation and to produce internalized fear by triggering a false memory. Optogenetics allows for selective inhibition of neural nucleus and it has been shown that the cerebral cortex is required for fear potentiation in startle protocol.[35] Further studies using optogenetics are promising to accurately investigate the neurobiology of the startle and its links with the neural pathways of emotional processing.

---

[31] Mathieu Nollet, Anne-Marie Le Guisquet, and Catherine Belzung, "Models of Depression: Unpredictable Chronic Mild Stress in Mice," in *Current Protocols in Pharmacology* (June 2013): Chapter 5, Unit 5.65. doi:https://doi.org/10.1002/0471141755.ph0565s61.

[32] Yann S. Mineur, Catherine Belzung, and Wim E. Crusio, "Effects of Unpredictable Chronic Mild Stress on Anxiety and Depression-like Behavior in Mice," in *Behavioural Brain Research* 175, no. 1 (November 25, 2006): 43–50. doi:https://doi.org/10.1016/j.bbr.2006.07.029

[33] Catherine Belzung, Marc Turiault, and Guy Griebel, "Optogenetics to Study the Circuits of Fear- and Depression-like Behaviors: A Critical Analysis," *in Pharmacology Biochemistry and Behavior* 122 (July 2014): 144–57. doi:https://doi.org/10.1016/j.pbb.2014.04.002

[34] Steve Ramirez, Xu Liu, Pei-Ann Lin, Junghyup Suh, Michele Pignatelli, Roger L. Redondo, Tomás J. Ryan, and Susumu Tonegawa, "Creating a False Memory in the Hippocampus," in *Science* 341, no. 6144 (July 26, 2013): 387–91. doi:https://doi.org/10.1126/science.1239073

[35] Aldis P. Weible, Christine Liu, Cristopher M. Niell, and Michael Wehr, "Auditory Cortex Is Required for Fear Potentiation of Gap Detection," in *The Journal of Neuroscience* 34, no. 46 (November 12, 2014): 15437–45. doi:https://doi.org/10.1523/JNEUROSCI.3408-14.2014

5 Conclusion

Animal and human models of startle highlight the strong link between startle and emotions. Indirect data suggest that startle is a necessary condition for emotions, although not sufficient. Startle is a critical feature for surprise and reveals the very basic emotional load within a more complex emotional processing. Through a neurobiological perspective, the startle reflex is both very simple and embedded within a wide system of interconnected neural pathways related to emotions. Through a first person perspective, startle may be related to surprise, itself embedded within larger emotional processes. An integrative understanding of emotions may require the investigation of the startle from a third person perspective with the experiences of emotion through a first person perspective.

Part III
Surprise and the Body

If the Body Is Part of Our Discourse, Why Not Let It Speak? Five Critical Perspectives

Maxine Sheets-Johnstone

Abstract Of the five perspectives set forth in this essay, four of them specify obstacles that block experiential understandings of emotions. The obstacles in one way and another subvert the living body, whether presenting it as a mere face or as an ahistorical adult body, as an embodied phenomenon or as a brain unattached to a whole-body nervous system. Such accounts bypass the affective dynamics that move through bodies and move them to move. Being true to the truths of experience, the fifth perspective, requires recognition of our infancy and even of our prenatal lives, both of which are tethered to developmental movement. It furthermore requires recognition of affective realties as subject-world relationships and recognition of the dynamic congruency of emotions and movement. In the end, the perspectives lead us to inquire about "the things themselves."

Keywords Animate · Dynamic · Brain · Embodied · Infancy

The five critical perspectives described and exemplified in this essay are: (1) the absence of the body below the neck; (2) the after-life of adults; (3) the lure of embodied simulation and microphrenological reductionism; (4) the whole-body nervous system: real-life, real-time embodiment; (5) the challenge of being true to the truths of affective experience. In its first four perspectives, this essay focuses on ways in which our understanding of emotions is cut short, that is, on obstacles to fathoming emotion and to carrying out veridical analyses of its experienced realities. In its fifth perspective, the essay shows in initial ways how, by enlisting other perspectives, we have the possibility of meeting the challenge of being true to the truths of affective experience and ultimately engage in phenomenological elucidations.

M. Sheets-Johnstone (✉)
University of Oregon, Eugene, OR, USA
e-mail: msj@uoregon.edu

N. Depraz, A. J. Steinbock (eds.), *Surprise: An Emotion?*, Contributions To Phenomenology 97, https://doi.org/10.1007/978-3-319-98657-9_5

1 The Absence of the Body Below the Neck

A provocative observation made in the course of a panel discussion on "Expression" during a conference on "Emotions Inside Out: 130 Years after Darwin's *The Expression of the Emotions in Man and Animals*" (Ekman et al. 2003; the conference proceedings were subsequently turned into a book) aptly captures the commonly truncated affective-kinetic character of adult emotion as rendered by science. An unidentified audience member comments, "I've been so excited by this whole presentation of this session [on "Expression"] because everybody is coordinated into one unit, but what has fascinated me is the absence of the body below the neck [laughter]" (ibid: 273). He or she goes on to explain: "I was fascinated by hearing the words, by seeing the faces, but I did not see the talking by the fingers, by the hands, by the movement, poise, and pattern of the people that were moving, sitting, or shifting." He/she then asks "if there is any further matter going on with the body as a Gestalt when you are communicating with your voice and your face" (ibid).

Well-known psychologist Paul Ekman responds first by citing the "pioneering work" of David Efron on gestures – what Efron refers to as "emblems" – and comments, "They are the only body language" (ibid: 273). But he goes on to mention Efron's specification of "speech illustrator movements," and then alludes to his own research on "bodily movements that we called *self-manipulative movements,*" movements such as playing with one's hands or scratching one's face (ibid: 274; italics in original). He concludes his response by stating, "There are other approaches that aren't looking in this formalistic way, but are looking at the flow, or quality, of movement," and goes on to remark, "These are people who primarily come out of dance. It doesn't appear that these body movements are as direct a signal source for emotion, in humans at least, as the face and voice. That's why we couldn't have found someone able to give a scientific talk on the body movements of emotion" (ibid).

The "absence of the body below the neck" might well be characterized as a chronic metaphysics of absence in "scientific talk on the body movements of emotion," and the absence of "flow, or quality, of movement" as a chronic absence of both dynamics and first-person experience in "scientific talk on the body movements of emotion." It is pertinent to point out with respect to both absences that phenomenological analyses are open to verification within a methodology no less demanding than that of Western science, which means they can be brought to self-evidence by anyone caring to examine experience. The absences readily conceal the qualitatively blinkered bodily life of most present-day scientists, perhaps even of most present-day philosophers, if not of academic and non-academic people generally. The deficiency of a blinkered science in particular notwithstanding, we can nonetheless clearly recognize that the absences constitute absences of a whole-body qualitative dynamics. Our voices and faces are part and parcel of those dynamics, part of the qualitative affective-kinetic dynamics created by otherwise spontaneously whole moving bodies. Animate bodies are indeed ones from which movement *flows*, and in flowing, creates a qualitative dynamic that can be elucidated in fine phenomenological detail (Sheets-Johnstone 1999a, b/exp. 2nd ed. 2011, 2009), a detailing that in truth is far more affectively and kinetically elucidating and exacting than scientific disquisitions on emblematic gestures, illustrator movements, and self-manipulative movement.

What indeed is "going on with the body as a Gestalt when you are communicating with your voice and with your face?"

If the body is part of our discourse, why not let it speak?

2 The After-Life of Adults

In his Introduction to *The Expression of the Emotions in Man and Animals*, Darwin voices concern that his analysis have a sound foundation and in this context calls attention straightaway to infants. He states: "In order to acquire as good a foundation as possible, and to ascertain, independently of common opinion, how far particular movements … are really expressive of certain states of the mind, I have found the following means the most serviceable. In the first place, to observe infants; for they exhibit many emotions, as Sir C. Bell remarks, 'with extraordinary force'; whereas, in after life, some of our expressions 'cease to have the pure and simple source from which they spring in infancy'" (Darwin 1965 [1872]: 13). What Darwin specifies as the "after life" of adults with respect to "the pure and simple source" of emotions in infancy can surely be described as a shift away from the animate body, in more precise terms, as an espousal of cognition, measured intellect, information-processing, the amygdala and other brain entities, and so on, over the realities of everyday spontaneous feeling in the form of affective-kinesthetic sentience – in short, a definitive predilection for mind over body in a literal spatial as well as ontological and epistemological sense.

It is relevant in this context to recall Jung's observation that "[e]motions are instinctive, involuntary reactions," and that "[a]ffects are not 'made' or wilfully produced; they simply happen" (Jung 1968, pp. 278–79). Emotions do indeed "spring" from the body and in their own distinctive qualitative kinetic dynamics, as both Darwin and Bell demonstrate graphically as well as descriptively (Darwin 1965 [1872], Bell 1844). In his final chapter on the expression of emotions titled "Concluding Remarks and Summary," Darwin emphasizes "the intimate relation which exists between almost all the emotions and their outward manifestations" (Darwin 1965 [1872]: 365). Indeed, both implicitly and explicitly throughout his text he has validated the intimate bond between movement and affect. Infant psychiatrist and clinical psychologist Daniel Stern aptly terms these kinetically-charged affective dynamics "vitality affects" (Stern 1985, 1993). Obviously, to appreciate them, we need to regain touch with our primordial animation by affectively and kinetically interrogating our "after life" as adults. When we do so, we are led to recognize the dynamic congruency of emotions and movement. In an article titled "Emotions and Movement: A Beginning Empirical-Phenomenological Analysis of Their Relationship" (Sheets-Johnstone 1999a), I spelled out that congruency, documenting its reality in scientific studies of animate life, i.e., in ethologist Konrad

Lorenz's description of fear in the readily notable behavior of a greylag goose; its reality in literary descriptions of people's reactions, i.e., in William Faulkner's vivid narrative of a character's fear in his novel *Sanctuary*; and its reality in an imagined scenario in which one is being pursued by an unidentified assailant at night in a deserted area of a city, i.e., in a phenomenological analysis of fear that sets forth its movement dynamics. In all three instances, we find a formally congruent dynamic; that is, emotions move us to move in ways that are dynamically congruent with the dynamics of the feelings we are experiencing. It is indeed because they are *formally* congruent that emotion and movement can be separated from each other, a separation that makes possible both the feigning of emotion and the restraining of its expression. In other words, we can go through the motions of a feeling, ostensibly showing others we feel a certain way, and we can restrain ourselves from moving in the way we feel inclined to move. In effect, we can smile when we actually feel no friendliness and we can refrain from a felt urge to slap someone in the face. It is important to point out that we know we can pretend or inhibit in a prereflective sense; our knowledge comes from our own lived experiences of the dynamic congruency of affect and movement. In particular, we have not pondered over the relationship between our feelings and our movement and come to realize that – "Hey! – I can pretend and I can inhibit!" Though we are typically if not certainly restrained in childhood from hitting or harming others, and thus inhibit our affective-kinetic proclivities, we are not typically apprised of the relationship between affect and movement and have not thought through on our own their relationship. Moreover no one actually teaches us to dissemble or to feign. In short, though emotions "simply happen," as Jung pointedly observes, we are free to follow through on their dynamics, moving in ways they move us to move or not. Similarly, we are free to feign their dynamics, moving or not moving in ways that are incongruent with our feelings. In our after-life as adults, we easily both feign and restrain. Bodily-felt feelings and bodily movement are commensurately narrowed. In our after-life as adults, we have a tendency not just to shut down the "pure and simple source" from which emotions spring in infancy, namely, our affectively felt bodies that move us to move, but to forget that source ever existed. In turn, we dampen our range of expressive movement and forget that we are, in Husserl's perspicuously exacting words, basically and quintessentially *animate* organisms. In turn, and in short, we cultivate an adultist perspective on emotions.

Again, it is appropriate to ask: If the body is part of our discourse, why not let it speak?

3 The Lure of "Embodied Simulation" and Microphrenological Reductionism

As if prescient of the direction in which his neuropsychiatric profession will take him, Edmund Jacobson comments in his 1970 book *Biology of Emotions* that "those who would do homage to the brain with its ten billion cell-amplifiers can well continue to do so," but they must also not overlook empirical evidence: that "muscles and brain proceed together in one effort-circuit, active or relaxed" (Jacobson 1970, pp. 36, 34). Through his clinical studies with trained observers, people who are not casual subjects as in traditional scientific studies, but trained in the technique of "auto-sensory observation" that he developed, Jacobson shows that "objective and subjective data," i.e., neuromuscular activity on the one hand, and attention, emotion, and imagination, for example, on the other, "indicates conclusively that when the trained observer relaxes the neuromuscular elements apparently specific in any mental activity, the mental activity as such disappears accordingly (ibid., p. 35). Note that Jacobson does not say that *all* mental activity disappears, only the mental activity ongoing before relaxation of neuromuscular elements.

Jacobson's empirical evidence of a singular muscle-brain 'effort-circuit' accords with Darwin's basic insight that movement and emotion go hand in hand. It accords as well with his seminal insight based on worldwide observations of animate life: "Experience shows the problem of the mind cannot be solved by attacking the citadel itself. – the mind is function of body. – we must bring some *stable* foundation to argue from" (Darwin 1987 [1838–1839]: 564; italics in original). Bodies are indeed the ground floor of animate life. Their kinetic/tactile-kinesthetic invariants – and proprioceptive corollaries thereof, e.g., the slit sensilla of spiders, the campaniform sensilla of insects – undergird the distinctive everyday affective-kinetic dispositions and capacities of each species of animate form. Morphology in the broad evolutionary sense of species-specific and species-overlapping kinetic/tactile-kinesthetic invariants thus provides precisely the stable foundation that Darwin specifies as necessary to veridical understandings of mind. Kinetic/tactile-kinesthetic invariants are the foundation of *synergies of meaningful movement* (Sheets-Johnstone 2009, 2010, 2011). Synergies of meaningful movement are to begin with *motivated* and remain *affectively charged*. They are thus affectively informed from beginning to end.

In contrast to Jacobson's empirically-grounded "muscle-brain effort circuit" and its resonance with Darwin's insights into emotion, movement, mind, and body, present-day neuroscience and cognitive science reduce the kinetic-affective-cognitive realities of animate life to neuronal activity in *the brain*. The lure of explaining ourselves to ourselves atomistically in this way has spread beyond the laboratories of neuroscience and the academic halls of cognitive science and philosophy, and other disciplines as well, to popular culture. Indeed it appears an irresistible lure, even to the point of absurdity. Consider, for example, an advertisement by The Teaching Company (*Science News* (vol. 175, no. 13, p. 3) that specifies a course taught by a neuroscience professor at a well-known and reputable university. The course is titled "How Your Brain Works." It is described as follows: "Everything you hear, feel, see, and think is controlled by your brain. It allows you to cope masterfully with your everyday environment and is capable of producing breathtaking athletic feats, sublime works of art, and profound scientific insights. But its most amazing achievement may be that it can understand itself" (vol. 175, no. 13, p. 3).

The brain is clearly the oracle at Delphi, the Mecca, so to speak, to which all questions about ourselves and our relationship to our surrounding world are addressed and from which all questions will in time be duly answered. In effect, the *fons et origo* of our effective relationships with other humans and with the array of objects we experience in the world about us, the fount of our knowledge of the world and of other people is neurological and lies in our own heads in the form of "internal non-linguistic 'representations' of the body-states associated with actions, emotions, and sensations." In the brochure summary of his guest lecture in conjunction with his Arnold Pfeffer 2009 prize award, Vittorio Gallese states just that. A critical problem lies in such a thesis. It is all but explicit in Gallese's own detailed account of "The Inner Sense of Action: Agency and Motor Representations" that appeared in an article published in 2000 (Gallese 2000). It is all but explicit in a different sense in his co-authored 2007 article with David Freedberg titled "Motion, Emotion and Empathy in Esthetic Experience" in which Freedberg and Gallese state that their purpose, is to "challenge the primacy of cognition in responses to art" and to propose instead that "a crucial element of esthetic response consists of the activation of embodied mechanisms encompassing the simulation of actions, emotions and corporeal sensation, and that these mechanisms are universal" (Freedberg and Gallese 2007, p. 197). In the first instance, it is a question of Gallese's underscoring of what he terms the "*common code*" (Gallese 200, p. 24; italics in original) of neuronal activity, namely, that all electromagnetic, mechanical, and chemical energy that constitutes stimuli of one kind and another, are "transduced" into action potentials. Action potentials of nerve cells are the common code of all stimuli. The singular problem is how this common code of all neural activity is itself transformed into "representations," a word that, incidentally, Gallese himself puts in quotation marks in speaking of "internal non-linguistic 'representations' of the body-states associated with actions, emotions, and sensations." How, in brief, can a common code – a same-for-all-stimuli – eventuate in distinctive representations – in a representation now of someone picking up a glass, in a representation now of horror in seeing a painting by Goya in which bodies are being mutilated, in a representation now of a tuber being "'not edible'" (p. 32)? The "common code" was recognized by Descartes and properly recognized precisely as a problem. Gallese's solution is to "naturalize" representation by giving it a home in the brain via "embodied simulation" (pp. 24). "[C]ausative properties," he states emphatically, "*are* content properties" (pp. 30, 34). In effect, as he goes on to state, "[T]o be phenomenally *conscious* of the meaning of a given object depends also on the *unconscious* simulation of actions directed to that object" (p. 32). Mirror neurons, as must be evident by now, are the key to naturalizing representation; they give representations substantive reality in the form of embodied simulations. The problem of explaining how a common code can eventuate in distinct representations is thereby elided. What Gallese terms "the safe bastions of physicalism" (p. 37) are indeed secured, for no further mention is made of a common code; it has been superceded by the capacity of neurons in the brain to simulate what Gallese terms "actions, emotions, and sensations." He justifies his naturalization of representation specifically to the point of adducing meaning, as when, in follow up to his statement that "causative properties *are* content properties," he adds, that "at least for *some* forms of mental content – their meaning is literally *constituted* by the way they are 'enacted' by a situated and functionally grounded organism" (p. 30). His neuronal formulation of meaning comes close to Evan Thompson's later claim that "[t]he nervous system … creates meaning".

In the second instance, it is a question of passing over the very bodily experiences that necessarily ground any so-called embodied simulations of "actions, emotions, and sensations" in the form of neural representations. In brief, Freedberg and Gallese's conception of aesthetic experience rests on ready-made mirror neurons. Indeed, in their eyes, we seem to be born with them. Precisely in this context, it is notable and of considerable interest to point out that in their initial experimental research to determine whether mirror neurons exist in humans, Rizzolatti, Fogassi, and Gallese – the major neuroscientists at Parma – did not utilize PET scan studies to determine neuronal activity in human brains; they utilized *neuromuscular* studies (see Sheets-Johnstone 2012 for a full analysis of mirror neuron research). They quite breezily pass over this fact in their special 2006 neuroscience article in *Scientific American* titled "Mirrors in the Mind," an article that gives a background history and summary of their findings to date. In particular, in describing their first experiments with human subjects in which the aim was to determine "whether a mirror neuron system also exists in humans," they state, "As volunteers observed an experimenter grasping objects or performing meaningless arm gestures, for example, increased neural activation in their hand and arm muscles that would be involved in the same movements suggested a mirror neuron response in the motor areas of their brains" (Rizzolatti et al. 2006, p. 58). The *suggestion* of "a mirror neuron response in the motor areas of their brains" and in consequence of a mirror neuron system in the human brain is clearly what is of moment to them. Accordingly, they pass over the *kinesthetically interesting finding* because of their desire to identify "the exact brain areas" that are activated when volunteers observe what they term "motor acts" (ibid.). Motorological talk readily and easily ignores kinesthetic experience in a way similar to the way in which talk of sensation readily and easily ignores dynamics, the experienced dynamics of both affect and movement. Real-life, real-time affective-kinesthetic experience is precisely what grounds the aesthetic experience of horror in face of a painting by Goya just as it grounds the everyday experience of horror in witnessing a freeway accident, videos of a tsunami or of a tornado, and so on. The experience of horror, here anchored in the recognition of extraordinary harm and pain is not located in canonical neurons – "premotor neurons" that are selectively activated according to the type of visual object being experienced. The experience is grounded in first-person tactile-kinesthetic/affective experiences, experiences we have all had in the course of growing up, of injuring ourselves in one way and another, in being injured by others, of being in pain from a stomachache, a toothache, and so on. As Sartre succinctly put it: "I live my body" (Sartre 1956, p. 325). I would add, I do not live my brain. Indeed, though it is distinctly a part of my body, my brain is nowhere in my experience, present not even as my heartbeat and breath are present.

Accordingly, if the body is part of our discourse, why not let it speak?

4 The Whole-Body Nervous System: Real-Life, Real-Time Embodiment

The sympathetic and parasympathetic nervous systems constitute the autonomic nervous system, a system that, along with the somatic nervous system, constitutes the peripheral nervous system. The peripheral nervous system is so named in relation to the brain and the spinal cord that are encased in the skull and vertebral column respectively, that is, in relation to the central nervous system. Evolutionary anatomist and former Alexander Agassiz Professor of Zoology at Harvard University, Alfred Romer, who wrote numerous, widely and well-used textbooks on evolutionary morphology, describes the sympathetic nervous system in general terms as follows: "Stimulation of true sympathetic nerves tends to increase the activity of the animal, speed up heart and circulation, slow down digestive processes, and, in general, to make it fit for fight or frolic" (Romer and Parsons 1977, p. 501). The parasympathetic often works in opposition to the sympathetic. Though it promotes digestion, for example, and other vegetative functions, it commonly constricts or contracts, and tends to slow down activity. Both sympathetic and parasympathetic are anatomically as well as functionally distinct parts of the autonomic nervous system. Parasympathetic fibers emerge from the cranial and sacral regions of the central nervous system, that is, from the brain and lower spinal cord. Topographically, the sympathetic system has its origin in the thoracic and lumbar regions of the central nervous system, hence its alternative name, the "thoracolumbar" system. Of particular interest is the fact that sympathetic fibers extend outward from each side of the spinal cord, and, unlike parasympathetic fibers, they synapse at ganglia at a distance from their target organ, be it heart, stomach, or intestines, for example, many of the ganglia forming what is called the sympathetic chain, a chain parallel to the vertebral column. The chain is an elongated synaptic junction, that is, a neural orchestrating column from which postganglionic fibers extend to internal organs – heart, bladder, liver, intestines, genitalia, stomach, and so on. Insofar as ganglia within the chain serve to mobilize organs, muscles, and other parts of the body – for example, bronchiole dilatation and dilatation of heart arteries – and to increase the rate of contraction of heart muscle and to contract the anal sphincter, for example, the sympathetic chain is obviously of moment in the day-to-day contingencies, opportunities, surprises, and challenges of animate life – in what Romer in general terms calls fight and frolic. In this respect, the sympathetic chain recalls a poem by a newspaper columnist, Bert L. Taylor, about the enlarged segment of the spinal cord of the dinosaur Diplodocus between its hips, an enlargement that notably exceeded the size of its brain (Romer 1954, vol. 1, p. 96):

Behold the mighty dinosaur,
Famous in pre-historic lore,
Not only for his power and strength
But for his intellectual length.
You will observe by these remains

The creature had two sets of brains–
One in his head (the usual place),
The other at his spinal base.
Thus he could reason *a priori*
As well as *a posteriori*.
No problem bothered him a bit;
He made both head and tail of it.
So wise was he, so wise and solemn,
Each thought filled just a spinal column.
If one brain found the pressure strong
It passed a few ideas along.
If something flipped his forward mind
'Twas rescued by the one behind.
And if in error he was caught
He had a saving afterthought.
As he thought twice before he spoke
He had no judgement to revoke.
Thus he could think without congestion
Upon both sides of every question.
Oh, gaze upon this model beast,
Defunct ten million years at least.

The point of the above quite summary account of the autonomic nervous system, particularly the sympathetic and the sympathetic chain – and the point of the poem and its emphasis on the perks of a double brain in a single body – is to bring into sharp focus the whole body nervous system, the whole body nervous system that includes not only the autonomic system, but the somatic system – sensory and kinetic – and the central nervous system as well. While it is fashionable in today's neuroscience to single out *the brain* – and fashionable in a near decerebrate fashion to do so – *the brain* is clearly and in actuality part and parcel of the whole body nervous system. Accordingly, if we talk in terms of embodiment and about entities or things being embodied, we should be speaking of the whole body nervous system: the whole body nervous system is what is embodied along with muscles, internal organs, blood vessels, and so on. It is indeed the whole body nervous system that is embodied, fully embodied. Accordingly, we should realize that we cannot properly or logically speak of body and brain as two distinct entities. By the same token, and as indicated, we cannot properly and logically speak of "embodied emotions" any more than we can properly and logically speak of "embodied language," "embodied self-awareness," "embodied experience," and of "embodied movement." We can in fact ask: can there be any such thing as disembodied fear, joy, or disgust – any more than there can be disembodied language, experience, self-awareness, and movement? Fear, joy, disgust, conviction, hesitancy, and so on – all are dynamically lived through affective bodily experiences (see Johnstone 2012). Each affect *moves through the body, and moves the body to move* along distinctive dynamic lines (Sheets-Johnstone 2006). Sympathetic and parasympathetic nervous systems are essential to these emotional dynamics. Short of them – literally short of them when cut off from everything below the cerebrum as in so much of present-day research on the brain – we would not experience an accelerated or decelerated heart rate, relaxed or widening eyes, not to mention erection and ejaculation. Short of an inherently integrated whole body nervous system, we would not experience grief, delight, hatred, terror, or any other emotion, neuroscientist Jaak Panksepp's reductionist scenarios notwithstanding. While certainly contributing to understandings of the neural architecture of emotions, Panksepp, perhaps as a result of the present climate of apotheosizing the brain, contributes also to an already astonishing number of experiential attributions to the brain. He writes, for example, "In my view, emotional feelings represent only one category of affects that brains experience" (Panksepp 2005, p. 162). Experiential ascriptions to the brain to the contrary, *the brain* – the notable and highly revered human one – is like the brain of any other animate form of life. It does not and cannot experience affects: *it does not and cannot feel.*

In sum, the autonomic nervous system is as vital to survival as the neuromuscular system, the hormonal system, the circulatory system, and all other anatomical, chemical, and physiological built-ins of animate life. Moreover everything works together in this human body of ours as in the bodies of all other forms of animate life. Humans are not different because of an outsized brain. In spite of what well-known neuroscientist Ramachandran writes in extolling prose about our impending brain-in-a-vat existence – "We are all slowly and imperceptibly approaching the brain in the vat scenario where all functions will be literally at your fingertips as you become dissolved in cyberspace" (Ramachandran 2006, p. 2)[1] – and in spite of what he writes in excessive prose about conscious experience and a sense of self – "we take for granted in these enlightened times" that "our conscious experience and sense of self is based entirely on the activity of a hundred billion bits of jelly – the neurons that constitute the brain " (ibid., p. 1) – in spite of all his adulatory prose, *the brain* – any brain – cannot compete with the whole-body nervous system that our phenomenal bodies embody and that is integral to the dynamically experienced whole-body affective realities of animate being and animate life. Thus we may again ask:

If the body is part of our discourse, why not let it speak?

5 The Challenge of Being True to the Truths of Affective Experience

In accordance with Darwin's perspicuous observation about infancy cited in the context of the second perspective, we would do well to begin at the beginning, that is, with ontogeny, if we are to be true to the truths of affective experience. The task is challenging because our adult vantage point is so seductive. We forget that we all came into a totally unfamiliar world; indeed in the beginning, we entered like aliens

[1] We might in fact ask: What's the point of bipedality if you cannot walk? If you are a brain in a vat, you don't have a leg to stand on.

on a foreign planet. We were not, however, totally foreign to ourselves, for we had spent some time moving ourselves. Starting around the 11th of our 42 weeks in the womb, we opened and closed our lips, we turned our head, we moved our hands and feet. Later, approximately halfway through those 42 weeks, our sucking and grasping reflexes begin to develop; we kicked our legs kick and waved our arms, and might have put a thumb in our mouth (Furuhjelm et al. 1977, pp. 95, 124). In short, our prenatal life was a life of developing movement. Moreover sound was not foreign to us. As medical researchers have noted, "The uterus is no silent, peaceful environment. The woman's pulse is constantly pounding; the placenta surges and murmurs; at times the woman moves abruptly, or speaks loudly" (ibid., p. 124). In effect, when we come into earthly life, we are like aliens on a foreign planet, but we are not dumb to touch, movement, and sound. Indeed, we are not awaiting consciousness as Daniel Dennett declared when he wrote, "[consciousness] arises when there is work for it to do, and the preeminent work of consciousness is dependent on sophisticated language-using activities" (Dennett 1983, p. 384).

Being true to the truths of affective experience requires attention not just to infancy but to basic animate realities such as the startle reflex, a fundamental reflex that psychologists Carney Landis and William Hunt studied in experimental fine detail across a range of subjects, including pathological and nonhuman subjects. The startle reflex is an affectively-charged movement pattern and clearly experienced as such, both when we ourselves are startled and when we see others startled. Landis and Hunt describe the movement pattern of the startle reflex "to be a general bodily flexion … which resembles a protective contraction or 'shrinking' of the individual" (Landis and Hunt 1939, p. 23). They detail the pattern specifically as follows: "blinking of the eyes, head movement forward, a characteristic facial expression, raising and drawing forward of the shoulders, abduction of the upper arms, bending of the elbows, pronation of the lower arms, flexion of the fingers, forward movement of the trunk, contraction of the abdomen, and bending of the knees" (ibid., p. 21). The pattern is clearly complex. Landis and Hunt note too the responsive character of the pattern, namely, that "[t]he response is very rapid and follows sudden, intense stimulation," and that the "basic reaction [is] not amenable to voluntary control, [and] is universal" (ibid.). Finally, they point out that four possible emotions are connected with startle: curiosity, fear, annoyance, and "overflow effects," the latter arising because, as they put it, "the primary response is not sufficient to resolve all the motor tensions aroused." In short, the startle reflex is a phylogenetic affective-kinetic reality that attests to an elemental, that is, *foundational subject-world relationship*. In doing so, it testifies directly and with incontestable animated immediacy to the superfluity of embodying, embedding, enacting, and extending "mental processes," as philosopher Mark Rowlands, for example, judges necessary in what is advertised as his "[investigations into] the conceptual foundations of the new science of the mind" (Massachusetts Institute of Technology Press 2011 brochure: see Rowlands 2010).

Finally, being true to the truths of affective experience requires recognition of Nina Bull's extended study of emotion via hypnosis, which showed that a body must be free in a neuromuscular sense in order to feel an emotion. In particular, if one is hypnotically locked into a certain emotion —say disgust – and then asked to express a certain other emotion, one is unable to do so. As one subject stated, "I reached for joy – but couldn't get it – so tense"; as another said, "I feel light – can't feel depression" (Bull 1951, pp. 84, 85). When it was said earlier that emotions move through the body and move us to move, it could have been added that they may also move us to be still, but not on that account to be simply a bundle of sensations. In contemplating a star-studded night sky, for example, and being stilled by a felt sense of grandeur, the felt sense is moving through us and continuing to move through us in our full-bodied presence to the spectacle before us. We are not moving about, but we are not wholly still either; we are not in a fixed *state*, but are held in what might be described as a dynamic rapture or wonder. Equally, when in everyday life we feel too tense to reach for joy or too light to feel depression, the tension or the lightness we feel is a complex affective-kinetic dynamic that moves through us and that continues to move through us, deterring us for the time from any other affective-kinetic dynamic. In short, to describe ourselves as simply having constricting sensations, for example, when we feel too tense to reach for joy and having weightless sensations when we feel too light to be depressed, or to describe ourselves more generally as being a "mass of tactile, labyrinthine and kinaesthetic data," as Merleau-Ponty does in the context of "catching space at its source" (Merleau-Ponty 1962, pp. 249, 243, respectively), is to ignore the dynamics of movement and affect and of their dynamic congruency. Indeed, were we not already *naturally* animated in a congruent dynamic of feeling, both affective and kinetic, we would be unable to feign an emotion. Such a talent testifies not to an embodied mind but to a mindful body, a body that in a substantive tactile-kinesthetic/affective sense knows itself and its possibilities (Sheets-Johnstone 1999a). Being true to the truths of affective experience requires being attuned to bodily dynamics, bodily dynamics over sensation, and by the same token, attuned to movement over bodily position. In broader terms, to be true to the truths of affective experience requires attendance to phylogenetic, ontogenetic, and experiential truths, all of which are rooted in and attest to the foundational animate realities of bodily life. Accordingly, should we not proceed 'to the things themselves', *to those otherwise eclipsed phenomenological realities of being animate organisms*? Should we not put a methodologically astute ear to our bodies and listen to their dynamics?

If the body is part of our discourse, why not let it speak?

References

Bell, Sir Charles. 1844. *The anatomy and philosophy of expression*. London: John Murray.

Bull, Nina. 1951. The attitude theory of emotion. In *Nervous and mental disease monographs*, x ed., x–x. New York: Coolidge Foundation.

Darwin, Charles. 1965 [1872]. *the expression of the emotions in man and animals*. Chicago: University of Chicago Press.

———. 1987. In *Charles Darwin's notebooks, 1836–1844*, ed. Paul H. Barrett, Peter J. Gautrey, Sandra Herbert, David Kohn, and Sydney Smith. Ithaca: Cornell University Press.

Dennett, Daniel. 1983. Intentional systems in cognitive ethology: The 'Panglossian' Paradigm Defended. *Behavioral and Brain SCIENCES* 6: 343–390.

Ekman, Paul, J. J. Campos, R. J. Davidson, F. B. M. de Waal. 2003. Expression: Panel discussion. In *Emotions inside out: 130 years after Darwin's The expression of the emotions in man and animals*, Annals of the New York Academy of Sciences. vol. 1000, 266–278. New York: New York Academy of Sciences.

Freedberg, David, and Vittorio Gallese. 2007. Motion, emotion and empathy in aesthetic experience. *Trends in Cognitive Science* 11 (5): 197–203.

Furuhjelm, Mirjam, Axel Ingelman-Sundberg, and Claes Wirsén. 1977. *A child is born*, rev. ed. New York: Delacorte Press/Seymour Lawrence.

Gallese, Vittorio. 2000. The inner sense of action: agency and motor representations. *Journal of Consciousness Studies* 7 (10): 23–40.

Jacobson, Edmund. 1970. *Modern treatment of tense patients*. Springfield: Charles C. Thomas.

Johnstone, Albert A. 2012. The deep bodily roots of emotion. *Husserl Studies* 28 (3): 179–200.

Jung, Carl. G. 1968. *The Archetypes and the Collective Unconscious*, 2nd ed. Trans. R. F. C. Hull. Bollingen Series XX. Princeton: Princeton University Press.

Landis, Carney, and William A. Hunt. 1939. *The startle pattern*. New York: Farrar & Rinehart.

Merleau-Ponty, Maurice. 1962. *Phenomenology of Perception*, Trans. Colin Smith. London: Routledge & Kegan Paul; 1945. *Phénoménologie de la perception*. Paris: Librairie Gallimard.

Panksepp, Jaak. 2005. Affective consciousness: Core emotional feelings in animals and humans. *Consciousness and Cognition* 14 (1): 30–80.

Ramachandran, V.S. 2006 (1.10.06). "Mirror Neurons and the Brain in the Vat" (Response to "What Is Your Dangerous Idea?"). *Edge: The Third Culture*.

Rizzolatti, Giacomo, Leonardo Fogassi, and Vittorio Gallese. 2006. Mirrors in the mind. *Scientific American* 295(November): 54–61.

Romer, Alfred S. 1954. *Man and the vertebrates*. Vol. 1. Baltimore: Penguin Books.

Romer, Alfred S., and Thomas S. Parsons. 1977. *The vertebrate body*. Philadelphia: W. B. Saunders Company.

Rowlands, Mark. 2010. *The new science of the mind: From extended mind to embodied phenomenology*. Cambridge: Bradford Books/MIT Press.

Sartre, Jean-Paul. 1956. *Being and Nothingness*. Trans. Hazel E. Barnes. New York: Philosophical Library.

Sheets-Johnstone, Maxine. 1999a. Emotions and movement: A beginning empirical-phenomenological analysis of their relationship. *Journal of Consciousness Studies* 6(11–12): 259–277; included in Sheets-Johnstone, Maxine. 2009. *The Corporeal Turn: An Interdisciplinary Reader*, 195–218. Exeter: Imprint Academic, Chapter VIII.

———. 1999b/exp. 2nd ed. 2011. *The primacy of movement*. Philadelphia/Amsterdam: John Benjamins publishing.

________. 2006. Sur la nature de la confiance. In *Les moments de la confiance,* ed. Albert Ogien and Louis Quéré, 23–41. Paris: Economica.

———. 2009. Animation: The fundamental, essential, and properly descriptive concept. *Continental Philosophy Review* 42: 375–400.

———. 2010. Kinesthetic experience: understanding movement inside and out. *Body, Movement and Dance in Psychotherapy* 5 (2): 111–127.

———. 2011. The corporeal turn: reflections on awareness and gnostic tactility and Kinaesthesia. *Journal of Consciousness Studies* 18 (7–8): 145–168.

———. 2012. Movement and mirror neurons: a challenging and choice conversation. *Phenomenology and the Cognitive Sciences* 11: 385–401.

Stern, Daniel N. 1985. *The interpersonal world of the infant: A view from psychoanalysis and Developmental Psychology*. New York: Basic Books.

———. 1993. The role of feelings for an interpersonal self. In *The perceived self: ecological and interpersonal sources of self-knowledge*, ed. Ulric Neisser. Cambridge: Cambridge University Press.

The Teaching Company. 2011. *Science News* 175 (13): 3.

Glancing at the Surface of Surprise

Edward S. Casey

> *There is always a world already interpreted, already organized in its basic relations, into which experience steps as something new, upsetting what has led our expectations and undergoing reorganizing itself in the upheaval.*
>
> –Hans-Georg Gadamer, *Philosophical Hermeneutics* (California 1976), p. 15

Abstract Glancing has a special affinity with surfaces, which draw its spontaneous attention with a special allure and force. What it picks up on surfaces – primarily visual ones but auditory and tactile as well (glancing occurs in aural as well as haptic modalities) – is often something it did not expect. When that happens, a unique kind of surprise is precipitated: however demure and gentle it may be, it may also illuminate existing puzzles as well as offer guideposts for further exploration. In this essay I investigate the triadic *conjunctio* between glance, surface, and surprise that deserves more phenomenological consideration than it has so far received. The direction of thought opened here is pursued in more detail in my book *The World at a Glance* (2007).

Keywords Glance · Surprise · Surface · Perception · Attention

1 Catching Sight of Surprise

To glance is to glance at a surface of some sort, however extensive or limited this may be. We glance at people, at physical objects, or events – at George, at a tree by the side of the road, and at a passing car. Each of these possesses a surface or set of surfaces at which we glance. As J.J. Gibson puts it, "the surface is where the action

E. S. Casey (✉)
Stony Brook University, Stony Brook, NY, USA

N. Depraz, A. J. Steinbock (eds.), *Surprise: An Emotion?*, Contributions To Phenomenology 97, https://doi.org/10.1007/978-3-319-98657-9_6

is."[1] It is what bears the looks and touches of everyday perception: attracts or repels them, but also supports them when they connect with it and linger. The support consists in being consistent enough, and lasting enough, to sustain these outreaches of the perceiving organism whether these occur by vision or by touch. Often opaque, and sometimes transparent, seen and felt surfaces are the ongoing partners of perception, that with which a glance or a touch colludes.

Behind every surface lurks a surprise – something that we did not suspect was there. In some cases not at all: this is the case of complete surprise, ranging from delight or chagrin to shock. In other cases, something else of which we had some sense that it might emerge but no conviction as to just when or just what it might be. Or else something just different from what we had expected would be the case – the surprise contravening what we thought was going to happen. Between these three basic kinds of surprise, there are many intermediary cases: degrees of surprise, or more exactly degrees within kinds of surprise. In the end, there are so many cases of surprise that our ongoing lives consist more in the surprising than in the expected. When the latter prevails, we are in the midst of routine, which consists in holding as much surprise as possible at bay so that we can deal with what life delivers in a maximally controlled and consistent way. But, as Gadamer intimates in the citation above, this cannot last: "something new" will "upset our expectations," and will itself have to undergo "reorganizing."

"Proximally and for the most part" (*zunächst und zumeist*: in Heidegger's recurrent phrase in *Being and Time*), we try to live a life in which the habitual prevails. Anything else is very uncomfortable if not intolerable. Thus it is not surprising that our bodily habitudes are at once massive and subtle – massively present in all we endeavor to do, while being subtly interwoven into virtually every daily activity. Merleau-Ponty's notion of "habit body" (*le corps habituel*) subtends everything we do in the daily course, so much so that we do not even notice how pervasive is its operation. And this is a good thing; otherwise, we'd be constantly re-orienting ourselves. The habitual is the basis not only of the overt "motility" (*mobilité*) of the lived body but of much else besides, including our characteristic ways of gesturing and touching, staring and glancing. The very *way* we look and feel, step and stride, and settle into a chair, is dependent on the body's commonly held habitudes.

Despite their alliance with the customary and the expected, habitual bodily movements are intimately related to innovative actions. Not just related to such actions but *making them possible*, for creative actions often presume a bodily-habitual basis. Only from such a basis – often consisting in specific skills that are essential to the accomplishment of these actions – does coherent experience happen. Here again, in very midst of such coherence and consistency, the surprising arises – this time not from behind sheer surfaces but from corporeal actions and movements themselves: as if it were an emergent property. When Picasso said that "je ne cherche pas, je trouve," he was referring to much the same phenomenon. Thanks to his immensely adroit skills as an artist, he did not need to search further at the level of

[1] J.J. Gibson, *An Ecological Approach to Visual Perception,* (Englewood Cliffs, N.J.: Erlbaum Associates, 1986), p. 46.

his habitual-bodily skills. By virtue of these same skills, he was able to *find* what came forth under his hands and eyes: to receive and accept it as such. This came as a surprise but not as a complete surprise. Or we can say that what was surprising came *right from* what Picasso had already incorporated at the deeply habitual level of his artistic activity. It came forward like a new aspect – or better, a new dimension or parameter of this same activity.

Stravinsky reports much the same situation. For him, only within the frame of established musical norms in a given tradition can innovation, including radical innovation, occur. As he writes in *The Poetics of Music*: "In art, as in everything else, one can build only upon a resisting foundation… My freedom thus consists in my moving about within the narrow frame that I have assigned myself."[2] Freedom in composition – which engenders surprise for the composer and listener alike – emerges only from familiarity with previous musical rules and styles that have become habitual: not merely in rejecting these customary practices, as in outright rebellion, but as a creative modification of them.

2 Surprise Arises from the Flesh

If this is true, can we then say that the mass of bodily habitudes is the equivalent of the set of surfaces with which observant perception has to do – given that both are accustomed citizens of our familiar life-world and that the surprising comes forth from both (and frequently from the interaction of each with the other)? But are they not irremissibly different? What allows them to interact in the generation of surprise? I suggest that Merleau-Ponty's later notion of *flesh* provides a deep linkage between two such otherwise different things as habits and surfaces. Just as flesh is a thick covering of the body, so the body's own habitual matrix forms a flesh-like aura around everyday activities – an aura that acts as a resource on which routine as well as innovative actions can draw. Similarly, there could be no perception of the life-world without dealing with the surfaces that surround us at all times: taking them in, judging their depth and opacity, in short witnessing them and then deciding (often quite spontaneously) what to do with or about them. These surfaces are in effect the flesh of things. With habits and surfaces alike, there is a sense of having to work *with* them and to work our way *through* them as if both were modes of flesh. The habitual and the surfacial alike have *épaisseur* in Merleau-Ponty's word: 'thick depth' as if this word can be translated.

The striking fact is that it is from this very density – often obdurate and opaque at first take – that the surprising in the form of the novel emerges. It does not come from thin air, from nothing or next to nothing. as is implied in certain theories of imagination as fashioning its contents from the sheerly fabricated or merely fictitious: from a process of "nihilation" in Sartre's term thanks to imagination's "thin

[2] Igor Stravinsky, *The Poetics of Music* (New York: Random House, 1960), p. 68.

autonomy."[3] Rather, the surprising as what we "could not imagine to be the case" rises before us from the surfaces of "the flesh of the world" in Merleau-Ponty's phrase. The surprising emerges from this flesh in the presence of our own flesh; it comes forth as something unlikely and unexpected, that which we truly did not think would arise from our habitual behaviors or from the ordinary surfaces that we continually confront. Or perhaps even *could not* arise from them as if it was truly *unimaginable*. In this latter case, the surprising verges on the *amazing* –as when we witness the tricks of a master magician or find ourselves astonished when listening to a jazz master like Keith Jarrett or a master rhetorician such as Martin Luther King, Jr.: how could such sounds or such words ever issue from such finite bodily movements?[4]

It is as if there were a common medium shared by habitual actions and the surfaces of surrounding things: both exhibit something consistently recognizable and obdurately repetitive within a thickly configured setting from which the surprising in its various avatars emerges.

3 On the Edge of the Sudden

When the surprising does appear, it appears *suddenly*. Such suddenness is this very feature that allows us to say that we are "caught by surprise." The very etymology of *sur-prise* indicates that in being surprised, we are *taken over* (*prise/sur*) by what appears before us. This basic action has two parts: we are overcome and we are caught off guard, and the former because of the latter. To be caught off guard is to be vulnerable to whatever comes upon us – to be subject to its taking us over, preoccupying us, and (in extreme cases) over-whelming us. But this double action is intensified when the experience occurs suddenly. How so?

The very character of the sudden upsets our ongoing sense of our experience as unfolding at its own leisurely pace in accordance with our current concerns and interests. The sudden cuts across this smooth coursing as if it were a transverse cut: a *Querschnitt* in Husserl's graphic term. "Suddenly" signifies coming from we know not exactly where – cutting across time's unfolding obliquely rather than forming part of the lateral flow. Proximally and for the most part, we count on this flow to sustain our lives from day to day and minute to minute. Yet this very regularity puts the flow at risk: at risk of being countered by something sudden: undercut or overcut by it, disturbing the consistency and regularity of the flow.

The sudden is not only a creature of temporal disordering. It also cuts across the regularity of surfaces in space. Surfaces act to cover, and in several ways: as adhering

[3] See *Imagining: A Phenomenological Study*, 2nd ed. (Bloomington: Indiana University Press, 2000), chapter eight.

[4] I have explored the differences between the amazing, the astonishing, the curious, and the wondrous in *The World at a Glance* (Bloomington: Indiana University Press, 2007), pp. 213–14, 220, 243–44.

to (as with skin as the surface of our lived bodies) or fitting loosely over (as with most items of clothing) or veiling (where the motive is often to conceal what lies underneath). Each of these forms of covering comes with its own kind of edge: skin is the very edge of the lived body itself, clothes create the socially presented edges of these same bodies, and veils possess their own edges in the form of folds.

Edges themselves act to conceal; they are, as it were, the last stage of covering that which lies under or behind them. But they also act to reveal – to show that of which they are the edge: as a certain kind of thing (e.g., a hand emerging from the cuff of a shirt) or an event (as when the motion of leg pants seen from afar discloses a person who is walking) or even a thought (by way of a furrowed brow).

Most important, however, is the fact that *every surface ends in an edge*.[5] As measuring its spatial extent, the edge is part of its very being, intrinsic to its presentation to us in perception. Even if we are not focusing on its exact edges, to look at a surface is to be aware of its edged character. This is true of the most colossal surfaces – say, those that covered the Reichstag or parts of Marin County in Jean-Claude's bold art actions -- as well as the most modest, as with a sleeping mask whose edges are etched on the face of the sleeper. Whatever the extent of the surface, it cannot appear unedged. J.J. Gibson's axiom "the surface is where most of the action is"[6] needs to be supplemented: *the action of the surface reaches its apogee at its edge.*

It is from behind the edge of a surface that the surprising appears with sudden force. Indeed, appearing this way increases the intensity of the surprise, since edges enact a drama of sudden disclosure. It is not surprising that edges can engender a special state that I have called "shaking at the edge."[7] Appearing suddenly in the midst of a surface is much less likely to occasion sheer surprise, much less anxiety – rather, intimations as to what is happening under the surface (as when we perceive a couple whom we thought were indifferent to each other making motions under a blanket that clearly suggest that sexual intimacy is happening). For the surprising thing or event is, by definition, *still under the surface* and is to this extent not fully apparent: not as the thing or event itself. But that which comes out from behind an edge *comes out as itself*, nakedly and unashamably, and is something we have to cope with directly, as such: as the thing itself!

The suddenly appearing and the edges of surfaces constitute two intensifying bases of the experience of surprise that is occasioned by the perception of surfaces.

---

[5] Even a Moebius strip *has* edges even if we don't cross *over* them when we draw a continuous line along the middle of the strip.

[6] J.J. Gibson, *The Ecological Approach to Visual Perception* p. 23.

[7] See my article of this title, "Shaking at the Edge," forthcoming in an volume edited by Michael Marder. Considerably modified, this has been incorporated into my book, *The World on Edge*, chapter ten.

4 Glancing at Surprise

There is a special kinship between the glance and surprise to which we must now attend. Time is once again the key. If not instantaneous, the surprising is not long-lasting: the longer it appears steadily before us the less surprising will it be. Integral to surprise is its short-lived appearance: hence the link between the surprising and the sudden. Among modes of looking, glancing is best equipped to deal with the brevity of the surprising, for it moves quickly among its objects, darting between them without lingering on any one of them for very long. Any such lingering soon turns into gazing or scrutinizing – quite different forms of looking, ones better suited to the aim of getting better acquainted with what appears to us to the point of its becoming something familiar. The surprising, however, is something distinctively *unfamiliar* and *unaccustomed*. As such, it calls for being just noticed, and this happens best by glancing at it.[8] Glancing is looking with a light touch, barely grazing what we look at, bouncing off it as soon as contact is made.

For the surprising to *surprise* us – to take us over by its unexpected character – it must come into our ken only briefly, like a meteorite that flashes in the upper atmosphere only to disappear the next moment. For us to apprehend it, a glance is both necessary (to notice it at all) and sufficient (no further looking is called for). Between the glance and the surprising there is an inner bond.

At the same time – and by this I mean *in the same moment* – this glance must alight on some surface for it to make connection with its intentional correlate. The surprising must appear on *some surface* or, still more effectively, at the edges of that surface. Even the meteorite flashes out from a background that is in effect a cosmic surface, and stars scintillate upon the canopy of the night-time sky. The surprising occurs on, or at the edge of, a given perceptual surface that is its effective ground: that *upon which* it appears. This upon-which (*das worin* in Heidegger's term) is a feature of the presentational surface of the surprising. It furnishes the glance with a place of attachment for its encounter with what surprises us. It is the flesh of things and events from which the surprising suddenly arises. This from-which of surprise complements the upon–which of the surface where the surprising appears.

This indicates that the spatial dynamics of the glance is two-fold, *from* and *upon*, just as its temporal dynamics is also two-fold, *along* (the stream of time) and *across* (that same stream). We here detect the infra-structures of an experience that we otherwise tend to dismiss as trivial or take for granted as obvious.

[8] For more discussion of this aspect of glancing, see *The World at a Glance,* Concluding Thoughts.

5 The Surface of Surprise Itself

The title of this short contribution is: "Glancing at the Surface of Surprise." Is there a distinctive *surface of surprise as such?* If so, it is paradoxical: since it is tempting to think of surprise as puncturing the customary surfaces of convention and routine: as destroying their placid surfaces. Yet we rarely consider its own surface. I shall end by maintaining that surprise has a special surface of its own, one that is already punctured. I have in mind a situation akin to that invoked by Yeats in his great poem "Sailing to Byzantium":

> … man is but a paltry thing,
> A tattered coat upon a stick, unless
> Soul claps its hands and sing, and louder sing
> For every tatter in its mortal dress,
> Nor is there singing school but studying
> Monuments of its own magnificence…

The surface of surprise is no smooth, untroubled plane. Thanks to the desultory and unexpected character of surprise, its own surface is torn up from the start – and is never mended. Surprise tears apart its own surface of presentation; if it did not, it would not come forward as surprise – neither now nor in the future. The garments of its own surface are in tatters, raw in their texture and ragged in their edges.

Surprise never settles down – never becomes smoothly predictable or reliable: if it did, it would no longer count as surprise. Its surface is *how* and *where* it appears. Not as an object but as an *event:* as something that happens to us in ways we did not expect and cannot fully anticipate.[9] We can only glance at it, giving it our momentary attention even as *it*, this event of surprise, is itself utterly transitory – is always already passing us by and passing away.

Nevertheless, it is from a fugitive event such as surprise provides that we construct our lives in space and time. We clap our hands and sing, creating momentary continuities in repeating bio-rhythms, forging habitudes in the midst of the evanescent. We bring about "monuments of our own magnificence" as creatures with histories and languages, forming ourselves as enfleshed in words that perdure and perceptions that enliven. From within these moving monuments we prepare ourselves – time and again, surface by surface – for the surprising that punctuates our otherwise all too regular lives. No "singing school" is needed to bring about this quasi-autonomous action of creation and re-creation: we do it ourselves, we do it with ourselves and with others, in complex congeries of habits in contact with surfaces in scenes of surprise that animate us ever more.

---

[9] See Jean-Luc Nancy, "The Surprise of the Event," in his *Being Singular Plural*, tr. R. Richardson & Anne O'Byrne (Palo Alto: Stanford University Press, 2000), pp. 159–176. I am speaking here not of the surprise of the event but of *the event of surprise*.

Emotions, Volitions and Motivational Dynamics

Pierre Livet

Abstract Emotions are usually supposed to be passive states and volitions active ones. But if one takes into account their dynamical aspect, they can also be considered as two kinds that belong to a more inclusive category, namely the category of motivational dynamics. Volition implies a conative and dynamical aspect and its dynamics is sustained and oriented by one's motivation. Emotions also are motivated dynamical processes. In what follows I will examine in more details and compare the processes of emotions and the processes of volitions as two kinds of motivational dynamics. Both emotions and volitions can be active either at the explicit level of consciousness or in an implicit way. I will examine how implicit or explicit volitions can trigger implicit or explicit emotions, how occurrent emotions -implicit and explicit ones- can trigger volitions, and how the different interactions between emotions and volitions involve different kinds of surprise.

Keywords Surprise · Emotions · Volitions · Motivational dynamics · Conative · Implicit · Explicit · Occurrent emotions

1 Emotions and Volitions as Motivational Dynamics

Emotions are usually supposed to be passive states and volitions active ones. But if one takes into account their dynamical aspect, they can also be considered as two kinds that belong to a more inclusive category, namely the category of motivational dynamics. Volition implies a conative and dynamical aspect and its dynamics is sustained and oriented by one's motivation. Emotions also are motivated dynamical processes. Let us assume that we feel an occurrent emotion (not a dispositional one or a mood) when we expect something and notice a change in our situation that is less favourable, more favourable, or simply saliently different from our expectation (Livet 2002, 2016). In these situations we do not need to be explicitly aware of our expectation: it may direct our perception and action in an implicit way. As we are

P. Livet (✉)
Aix-Marseille University, Aix and Marseille, France
e-mail: pierre.livet@univ-amu.fr

N. Depraz, A. J. Steinbock (eds.), *Surprise: An Emotion?*, Contributions To Phenomenology 97, https://doi.org/10.1007/978-3-319-98657-9_7

not necessarily explicitly conscious of our expectation, the differential between our expectation and the change of the situation does not necessarily trigger an explicit and overt emotion of surprise. But there is at least an element of covert or implicit aspect of "surprise" in emotion, as we appraise a change or a differential of the situation relatively to our current expectation, even if we do not consciously categorize our emotion as "surprise".[1] Emotions are related to expectations that are not only cognitive, but also linked to the desire and possible actions that induce selective orientations of our perception: they are related to our motivations, and have to be analysed as one kind of motivational dynamics.

In what follows I will examine in more details and compare the processes of emotions and the processes of volitions as two kinds of motivational dynamics. Both emotions and volitions can be active either at the explicit level of consciousness or in an implicit way. I will examine how implicit or explicit volitions can trigger implicit or explicit emotions, how occurrent emotions -implicit and explicit ones- can trigger volitions, and how the different interactions between emotions and volitions involve different kinds of surprise.

If emotions and volitions are two kinds of motivational dynamics, we have to question our tendency to categorize emotions as passive, in opposition to volitions. In order to question – but also to understand- this tendency, we have not only to consider our internal dynamics, but also its relations to external dynamics, the dynamics of our environment. Volition is usually reduced to our "internal" dynamics: the combination of desiring to reach some goal and triggering as well as monitoring action in order to reach it. The triggering and monitoring processes are for a part dependent on the environmental dynamics: we perceive either an opportunity for satisfying a desire, or an obstacle to its satisfaction – in the latter case, action is needed in order to remove the obstacle. In the same way, our way of controlling our action is related to the reaction of the environment to the impact of our action. Of course our perception and our monitoring select the properties – and changes- of the environment in accordance with our subjective interests and concerns. Nevertheless, the selected properties are most of the time grounded in real dynamics of the environment. Volition implies the combination and articulation of our internal dynamics and the environmental one. In a sense, volition implies not only a pro-active aspect but also a "passive" aspect and reactive one, as it is receptive to the external dynamics. This passive aspect is amended by the selective aspect of our perception and appraisal and the part of our monitoring of the action that are both inspired by motivation.

Emotions imply also relations between internal motivations and external dynamics. They are related with our expectations; their motivational part selects and guides our perceptive anticipations and the evaluation or appraisal of the change of the environment – its external dynamics. We can also monitor in some way most of our

[1] This differential between our current expectations and some aspect of the situation is a common property of the burst of surprise and the burst of different kinds of occurrent emotions, even if emotions other than surprise do not imply an emotion of surprise as a component - this would lead to an infinite regression.

emotions – except traumatic or irrepressible ones. Emotions like joy, sadness, fear and anger may induce actions, and they even can be said to monitor actions, or at least our degree of subjective involvement into actions.

Frijda has noticed these relations between emotions and actions. He has defined emotion as "action readiness potential" (Frijda 1986). This definition raises two difficulties. First, there can trivially be action readiness potentials without emotion: we are prone to turn the handle of a door without any emotion. On the contrary we can feel emotions without "action readiness", as when we enjoy contemplating a beautiful landscape. Contemplation may imply attention, which is a component of action, but we could even enjoy contemplating without focalised attention, for example in mindfulness (Depraz 2014). Second, if emotions are action readiness "potentials", volition has to be classically defined as moving from a potential to an actual action (Deonna and Teroni 2012). But the transition from potential to actual remains to be explained, and as emotion are on the potential side and volition on the actual side, the gap between the two is an obstacle to the understanding of more entangled aspects of emotions and volitions as motivational dynamics.

Emphasizing the similarity and possible entanglement of emotions and volitions raises another objection: how can it help us to define the difference between emotions and volitions? In order to answer this objection, I have first to examine in more details our different experiences of volition, our different experiences of emotions, and the relations between these different kinds of experience.

2 Explicit and Implicit Emotions and Volitions

We have seen that we can distinguish different kinds of emotional experience: occurrent emotions and dispositional emotions. Occurrent emotion is related with conscious experience of feeling and emotion – even if sometimes it is difficult to categorize it. But we experience other kinds of emotions: we can behave as an angry person, while not consciously feeling anger. In this case, anger is only a dispositional emotion. In the same way, our volitions can be "occurrent" or not. In occurrent volitions, we are conscious of being engaged in volition. We could say that occurrent volitions are "explicit" volitions. As the notion of disposition is a bit tricky in the case of volition, I will use the term "implicit" instead of "dispositional". In the case of emotions, the opposition between implicit and explicit is not always equivalent to the opposition between occurrent and dispositional, for example when we feel an occurrent emotion but cannot "explicitly" put a name on it. Nevertheless, for my present purpose I prefer to use the distinction between implicit and explicit, as it is closer to the distinction between covert and overt consciousness than the opposition between actual and dispositional.

Most of the time, our usual actions are not accompanied by experiences of occurrent or explicit volitions. Usual actions are for the main part routinized forms of actions, triggered by affordances of our environment and motivated by our usual needs. Volitions are only implicit in these situations. They are related to "pro-tensions" – in Husserl's vocabulary-which are immediate anticipations. Nevertheless, neuroscientists have shown that volition is not the simple "act" of launching an action. One reason of giving up this simple version has been given by experiments on mirror neurons. Even if the powers of mirror neurons have been a bit overestimated, the interesting fact is that they are activated just by seeing the movement of another people - mainly in an action oriented towards a goal. In some pathology, people who see other's volitional movements have difficulties to inhibit similar movements. Seeing other's movement seems to imply a degenerative kind of implicit volition (and in this case, a truly dispositional one). We can infer that in non-pathological cases, normal people that see others' actions have to inhibit these tendencies to act. People with frontal deficits have also difficulties to inhibit actions that have to be postponed. This implies that in these later cases also we have to inhibit this kind of inhibition in order to trigger our own actions. Of course, most of the time we are not seeing other's movements and inhibiting the inhibition of a similar movement. But our current activities have inhibitory effects on the future action, and we have to inhibit these inhibitory effects in order to trigger the new action. Inhibiting the inhibition of the planned movement or its inhibition by other movements gives us an implicit feeling of volition, implicit inasmuch it gives rise only to feeling the movement as "our" movement and not to explicit consciousness of volition.

In order to become explicitly conscious of our volition, additional conditions are required. We are not explicitly conscious of our volition when we act because we are attracted by an affordance that triggers a routine. How can explicit consciousness of volition be triggered? I try to give an approximate (and not exhaustive) list of conditions.

1. We have to shift from one affordance to another one, maybe because one routine triggers another routine, or because we trigger two routines in the wrong order.
2. We have to combine two or more routines.

The combination of routines could explain Libet's experimental results. He asked participants to press a button, and at the same time to notice the position of the hand of a large clock at the moment in which they become aware of their act of volition. Participants notice the position of the clock a bit after having press the button. It seems strange, but do not forget that the instructions enjoin them to combine two actions, pressing the button and noticing the position of the clock. They have not only to inhibit the inhibition of pressing the button, but also to activate our recording of the position, and their experience of the triggering time of volition is not related with the simple action of pressing the button, but with the combination of the two routines.

3. We become conscious of our volition when carrying out a task needs effort and unusual concentration that are difficult to maintain, or when we have to look for a new idea, image or formulation, and have to reject irrelevant ideas and find new combinations.

4. We become conscious of our volition when we have to inhibit the attraction of an affordance of our environment or interrupt an attractive activity because our agenda requires that we have to shift to another less exciting one.
5. We become conscious of our volition when we hesitate between different choices –each of which satisfies a different criterion- and decide to select at the end only one of these alternatives.

To summarize, we become explicitly conscious of our volition when our motivational dynamics is not straightforward, is moved by different internal impulses or attracted by different affordances, or encounters the resistance of external dynamics. In other words, explicitly conscious volition is related to specific conditions (possible bifurcations or combinations, resistance, suboptimal attraction) of the relation between internal motivational dynamics and external environmental dynamics, conditions that transform the direct relation of the dynamics of action to its goal into an indirect and trickier relation.

In the emotional domain, we find similar patterns. What are "implicit" emotions? We can use this term in order to refer to motivational dynamics that are basically related with the pro-tensions and anticipations linked with to usual routines and make us sensitive to aspects of external dynamics that are relevant to our motivations. The external dynamics does not necessarily exhibit a noticeable differential with our expectations. These implicit feelings make manifest the interaction between our pro-tensions, anticipations, affective attitudes, and the positive or negative affordances of the environmental dynamics. We feel happy, sad, anxious, or find the situation tedious, but in these cases our feelings are more similar to the spreading of a mood than to the burst of an occurrent emotion.

We are explicitly conscious of occurrent emotions. This explicit awareness is triggered by the differential between our rather routinized expectations and the external dynamics. This differential implies a kind of surprise. Nevertheless, we feel surprise explicitly only when the differential is very strong, or when we were involved in an exploratory cognitive task and find unexpected things. We do not feel surprise explicitly when we feel fear, anger, or joy: the positive or negative valence of these emotions overwhelms the surprise related to the unexpected aspect of the external dynamics. In the same way, we do not explicitly feel surprise when we suddenly feel depressed because of a hopeless situation.

3 Similarity and Differences Between Emotions and Volitions

Situations in which we are explicitly conscious of volitions may induce us to feel occurrent emotions. Shifting from one affordance to another one may trigger emotions, inasmuch the new affordance is more attractive, or less attractive, or even repulsive. Combination of different routines may raise excitement. The effort that is needed to overcome an obstacle may raise a kind of anger, or pride if we overcome it. Inhibiting the attraction of an affordance may raise sadness, or pride in having been successful. Hesitation may raise anxiety and decision may raise relief –or regret. Occurrent emotions are strongly related to a variety of situations in which we are conscious of volitions, even if these different situations may raise opposite emotions.

All these reflections seem to weaken the usual opposition between emotions as reactive and volitions as pro-active, and to emphasize their similarities, due to their common status of motivational dynamics. We have to acknowledge that as long as we focus only on the internal dynamics volitions and emotions present similar features. We find a similar difference and articulation between implicit and explicit, or covert and overt consciousness of these two kinds of experience. If emotions are triggered by a differential between our anticipation and the present situation, volitions are related to differences between the present situation and our anticipation. One could object that volitions come from inside and emotions from outside. But volitions can also be triggered by external affordances, and emotions by internal reflections that lead us to self-evaluations.

Of course there are differences between emotions and volitions. One might refer to Searle's notion of directions of fit: for example, as beliefs in our mind are supposed to be representations of the world, the direction of fit of beliefs is from mind to world; when we want to satisfy a desire, we want to change the world in a way that satisfies our desire, and the direction of fit is from world to mind. According to his framework, emotions would be assumed to have mind-to-world direction of fit and volitions to have a world-to-mind direction of fit.

This is disputable. Volitions depend on our abilities – we can desire something that we are not able to get and to do, while we cannot experience volition about something that we cannot do- but the possibility of doing something depends also on the constraints imposed by reality. Emotions take into account the changes of the world, but they also depend on our expectations and desires.

Maybe the reason of these difficulties is that the dualistic relation between mind and world is a static one, and does not take into account the fact that the interaction is not an interaction between two states, but an interaction between two dynamics – in which the effect of one dynamics is modulated by the dynamics with which it interacts. In order to take into account this interactional relation, we need to make a step outside of the internal dynamics ("the mind") and consider its relation with external dynamics ("the world") as well as the way by which the external dynamics is modulated by the internal dynamics. Both volition and emotion imply a difference between internal and external dynamics, and in both cases, the external dynamics is appraised relatively to the internal motivational dynamics that modulates its effect. But the modes of modulation differ.

Let us try an analogy, and consider the modulation of the external dynamics by the internal one as a relation between an operation and its operanda, or a function and its variables and parameters (parameters are constants that enter into the calculus of the function; they can be chosen and adjusted in order to ensure that the results of the calculus of the function fit the data). In the case of volitions, the project of action depends mainly of the ability of the organism to act. The external dynamics do not change this ability. The constraints of the current state of the world modulate the actualisation of the action program and adjust the parameters of the function -the projected action. In the case of emotions, the external dynamics transforms the internal dynamics itself. For example it may induce a shift from an indifferent mood to a happy one, a transformation that is analogue to the shift from a linear function to a non-linear one –if we consider joy as analogous to acceleration or sadness to deceleration. In the case of volition, the impact of the external dynamics is monitored by the project or potential of the internal dynamics. In the case of emotions, the impact of the external dynamics modulates the pro-tensions, the expectations and the activities of the internal dynamics.

The fact that volitions and emotions can be either implicit or explicit implies that we can distinguish two sub-kinds of the emotional or volitional modulations of the external dynamics by the internal one. Emotions are "implicit" when we are not necessarily conscious of them as specific types of affects or affective bursts. Volitions are implicit when we act without an explicit consciousness of wanting our action.

Implicit volitions, like any kind of volitions, are submitted to the constraints of the situation; the parameters of our action are slightly adjusted in accordance with these constraints, but the adjustment does not imply a great revision of the parameters. Some affordance triggers a routine of action and inhibits the inhibition of action; the adjustment of the parameters is made "on line", without needing a change of the basic program of the routine.

In implicit emotions, the interaction between the external dynamics and its appraisal with respect to the internal ones determines the affective function, but once determined, the emotion is stationary. In a sense, in implicit emotions and volitions the influence of the external dynamics is at first dominant, as it gives its orientation to the internal dynamics, but after that it does not change the course of the internal dynamics.

4 Interactions Between Emotions and Volitions

Explicit volitions implies that external dynamics suggest to give up a routine and to change the process of action, or, on the contrary, impose obstacles to our action and constrain us to strengthen it in order to resist this pressure. Such incentive to change or resist modifies the internal dynamics. We have seen that such modification of the internal dynamics can trigger an emotion. Conversely, occurrent emotions can trigger actions – if we have in our repertory action programs that are appropriate as reactions to the change imposed by the external dynamics.

These remarks lead me to analyse the interactions between emotions and volitions while taking at the same time into account the differences between explicit emotions and volitions. As we have just seen, the conditions that are incentives for explicit volitions are also conditions for the activation of emotions. Most of the time, these emotions are implicit ones – our affective dynamics change- but we need not to be explicitly conscious of the category of our emotion. In these cases, explicit volitions trigger implicit emotions. Conversely, occurrent and explicit emotions trigger modifications of our conative attitudes, including volitions. In some cases, explicit emotions trigger only implicit volitions. In these cases, occurrent emotions directly trigger the onset of action – running away when frightened, attacking when angry- which most of the time implies only an implicit volition: we are not explicitly conscious of our volition as we are overwhelmed by emotion. Here we have symmetrical basic relations. In the first one, the flow of dynamics goes from explicit emotions to implicit volitions, in the second one, from explicit volitions to implicit emotions. I postpone the examination of the other two possible flows, from implicit volitions to explicit emotions and from implicit emotions to explicit volitions, until the examination of the different kinds of surprise related to all these relations.

In order to analyse the possible relations between implicit volitions and implicit emotions we need to consider in more details the process of inhibition- that we have at first detected in the case of volitions, which imply to inhibit the inhibition of the action program. Let us examine the case of a dynamics that flows from implicit volitions to implicit emotions. Affordances may directly trigger this inhibition of inhibition and the execution of the action program when the action is a routine (turning the handle of the door in order to come into the room). Achieving the action program without troubles, in a fluent way, is accompanied by an affect – feeling easiness- which is an implicit emotion. Feeling easiness is a relatively stationary affective dynamics that could be called an emotional or affective "attitude". Most of the times, such stationary affective attitude inhibits other implicit emotions – for example small stress or anxiety. Such inhibition is temporary and its effect may be weak. But it allows us to notice a similarity between the inhibition of inhibition of a specific action, which is related to implicit volition, and the inhibition of other emotional attitudes than the stationary one, which is related to our feelings during the action.

Let us consider now the relations between explicit emotions and explicit volitions (in this temporal order). Here occurrent emotions, like fear, anger, but also joy, inhibit the current routine and trigger new action dynamics. We have already examined the case in which fear triggers flight, but it was a case of implicit volition. This volition could become an explicit one when we have time before action and can at least briefly explore what possible action or reaction is better- for example, in case of fear, is it better to flight or to react aggressively; in case of joy, is it better to jump and dance or to stay at rest and savour the situation? In these situations, volition does not only imply double inhibition (inhibiting the inhibition of a routine action) but a triple one (inhibiting the double inhibition that triggers action). Occurrent emotion has done the first part of the job: inhibiting the current routine, and volition does the other part: inhibiting other possible actions for a while. The very volition that suspends action in order to explore other possible actions than the previous one is an explicit volition. We feel also explicit volition when we maintain the previous action despite the pressure of explicit emotions. In this case, we inhibit the inhibition of this action, a kind of quadruple inhibition!

5 Surprise and Entanglements Between Volitions and Emotions

I will now examine the eight possible combinations between explicit or implicit emotions or volitions, and the order of occurrence of emotions and volitions, and show how each of these combinations implies a different kind of surprise – in some cases surprise is reduced to a change of the affective tone or a change of intensity of involvement in action.

When an explicit volition (changing the type of action or resisting pressure) triggers only an implicit emotion, surprise is a cognitive feeling that becomes associated with the explicit volition when we discover that we need to change our routine in order to overcome an obstacle, or have to resist a pressure that was not anticipated. The implicit emotion itself (stress, irritation, etc.) does not contain surprise, but only a change of the affective state.

An implicit emotion can induce an explicit volition in a rather similar way: we are irritated by the current situation but are not explicitly conscious of our feeling. This irritation leads us to react suddenly, but we become explicitly conscious of this volition without being aware of its relation with our previous irritation. In this case, our emotion itself does not imply surprise, but we are surprised by the burst of this volition and action.

When an explicit emotion (fear for example) triggers an implicit volition (taking flight), surprise is a salient component of emotion. But like volition, which is not detached from the emotional pressure to take flight, surprise is not detached from fear.

An implicit volition can be one of the factors of an explicit emotion: we are doing a routine action and encounter resistance. This resistance raises an explicit emotion and an explicit surprise, related to the differential between the expectation that is embedded in our volition and the surprising resistance.

When an implicit volition (triggered by an affordance) is accompanied by the implicit emotion of easiness, we are not surprised, but the affect of easiness can inhibit other previous affects: the surprise is reduced to an evolution of the affects of the motivational dynamics.

An implicit emotion can trigger an implicit volition: most of the time, people that are stressed are not explicitly conscious of their emotion, but stress may have effects on volition – either reducing voluntary investment in action or on the contrary increasing it. Here again, there is no surprise, but only a change in the intensity of volitional involvement.

When explicit emotion induces explicit volition – when we do not immediately react to a change of the situation, but explore and evaluate different possibilities of reaction- this phase of reflective volition is also a way to assimilate the novelty of the situation and to progressively reduce or overcome the surprise triggered by the emotion.

An explicit volition can induce an explicit emotion: to be involved in a difficult or tricky action may induce both reinforcement of the intensity of volition and explicit emotions of excitement or anxiety. Here surprise can be a component of the volitional attitude and of the emotional one, but it will be more likely an explicit component of the volition, and an implicit component of excitement or anxiety.

To summarize, surprise is reduced to change of affects or changes of intensity when implicit volition triggers implicit emotions, and when implicit emotion induces implicit emotion.

Surprise is mainly cognitive, explicit, and related to the conditions of volition (resistance to a routine) when explicit volition induces implicit emotion. Surprise is also related to volition when implicit emotion induces explicit volition, but here surprise (a reaction to our sudden burst of action) is not only cognitive, but also affective.

Surprise is a salient component of emotion, but it is integrated into the affective differential, when explicit emotion induces implicit volition. Surprise has roughly the same role when explicit emotion induces explicit volition, but here reflective volition has the task of assimilating and reducing the surprise.

Surprise is present both in volition and emotion when explicit volition induces explicit emotion, but it is more explicit in volition that in emotion.

I have just given a sketchy description of some interactions between emotions and volitions when they are considered as two motivational dynamics. But I hope that the reader will be convinced that the analysis of emotions – particularly the analysis of their "surprise"' component- could be enriched by examining them in relations with volitions. In this domain of investigation, it would be fruitful to develop the studies of the entanglements between the different aspects of motivational experience (including not only volitions but also desires, efforts, moods, etc.), without taking for granted restrictive definitions that overestimate their oppositions and underestimate their intimate relations.

References

Deonna, J.A., and F. Teroni. 2012. *The emotions, A philosophical introduction*. London/New York: Routledge.

Depraz, N. 2014. *Attention et Vigilance*. Paris: Presses Universitaires de France.

Frijda, N. 1986. *The Emotions*. Cambridge: Cambridge University Press.

Livet, P. 2016. Emotions, beliefs and revisions. *Emotion Review* 8(3, July): 240–249.

———. 2002. *Emotions et rationalité morale*. Paris: PUF.

Part IV
Surprise in Hermeneutics

Call and Conversion on the Road to Damascus. Contributions to a Hermeneutics of Surprise

Jeffrey Bloechl

Abstract The experience of surprise involves an encounter with novelty that transforms one's understanding of the prevailing context and even of oneself, as a participant there. However, in the order of understanding absolute novelty is unintelligible, and absolute transformation would suspend the unity of the one who lives through it. We find indications for these theses in some reflection on Paul of Tarsus.

Paul was undoubtedly taken by powerful surprise when on the road to Damascus he heard the voice of Jesus. Yet what we know of his religious practice as an observant Jew suggests a range of conditions that may well have prepared him for what transpired there, and what we know of his subsequent actions suggest a long period of discernment. These considerations outline the possibility of further existential psychological work, without claiming to render Paul's experience fully transparent.

Keywords Conversion · Call · Paul · Revelation · Faith

1 Toward a Hermeneutics of Surprise

For many of us, the phenomenon of surprise brings to mind the image of sudden interruption, discontinuity and even reversal of previous conditions. Of course, often enough this does have a measure of plausibility, at least if we restrict ourselves to the *hic et nunc* of the experience itself. Whereas previously something was absent or unknown to me, it is now present before me. Or whereas previously things appeared to me in a particular way, I am made to see that in fact things are otherwise. This can provoke a sense of dread, as when a new realization undermines an entire worldview, but it can also be invigorating. The former possibility has become well known to us through the work of existential psychiatry and psychology. We find an interesting

J. Bloechl (✉)
Boston College, Chestnut Hill, MA, USA
e-mail: bloechl@bc.edu

N. Depraz, A. J. Steinbock (eds.), *Surprise: An Emotion?*, Contributions To Phenomenology 97, https://doi.org/10.1007/978-3-319-98657-9_8

study of the latter possibility in some passages of Henri Bergson's book on laughter.[1] Applying himself to the sort of stage comedy in which assumed conditions and familiar movements suddenly fail (the clown stumbles over a chair that we now realize has been there all along, etc.), Bergson suggests that the true source of our laughter is found in a sense of escape from the impersonal objectivity that has just been pierced. We are not mechanisms ordered to fixed patterns that regulate our movement, but rather living beings whose every movement is first a free creativity, *and in our laughter we give expression to a deep sense of this condition.* As a discrete event, what surprises does so by disrupting a range or sequence of repeatable events. It explodes the ordered pattern with all the force of an element or atom that does not belong there. Whereupon—let us follow Bergson's thinking through to the end – the inward shock that constitutes our feeling of surprise is quickly diminished in a return to what we experience as "the familiar."[2] After all, our feeling of surprise often quickly passes, and indeed few jokes to none can be repeated without losing their original force. They become, as I have just suggested, part of the fabric of the familiar.

Now the appearance of a dimension of familiarity in the experience of surprise introduces an important complication for the present investigation. Whereas there is only *equivocation* in our everyday use of the same word "surprise" to designate both an event and a response to an event, there is genuine *ambiguity* about the function and meaning of familiarity in relation to the disruption that surprises. The equivocation is easily dissolved by distinguishing between an event that surprises and the feeling of surprise that arises in response to it. And this serves to highlight the ambiguity. In fact, it is twofold. (1) To begin with, it is not immediately clear whether the general familiarity that is interrupted by the event that surprises us is therefore closed from that event or only contrasts with it. We might simply hesitate before the idea that a surprising event is strictly opposed to all previous conditions, without any continuity with them. And we might anticipate that in the absence of any continuity whatsoever, the event that surprises must be meaningless. Such a concern hangs most prominently over the entire philosophy of Levinas, for whom the face of the other person is pure self-expression, arriving from beyond any horizon. The face is thus a "shock" and even a "trauma" by which the very subjectivity of the subject is thrown back on itself to a point from which either of only two general possibilities arise: One might be converted from being for-oneself into being one-for-the-other, or one might recede back into insistence on oneself and, indeed, the comforts of the familiar. Without venturing into a detailed interrogation of Levinas's grounds for these claims, let us only note that an argument organized to defend the absolute otherness of the other requires the thought of an event that cannot be understood by the subject except at cost of already falsifying what gives itself to be seen. This can only mean that when we experience some surprise, we are, as subjects, already beginning to forget what surprises us. And in turn, this severe claim plainly implies

[1] H. Bergson, *Laughter. An Essay on the Meaning of the Comic* (Wilmington, DE: Dover Publications, 2005). I leave aside the plainly anti-Cartesian bent of Bergson's analysis.

[2] H. Bergson, *Matter and Memory* (New York: Zone, 1994), p. 253, note 23, citing with appreciation Alfred Fouillée's *Psychologie des idées forces* (1893).

a reduction of the life of the subject from *ipse* to *idem* that, as some of Ricoeur's later work reminds us, we ought not concede too easily.[3]

(2) A second ambiguity lies in the difference between thinking that the familiar is essentially a projection of the subject, as it seeks to make itself at home in the world, and thinking that to some degree the familiar arises from the manner in which the world, as it were, lends itself to the givenness of all data to consciousness. Here we might ask whether the very notion of surprise presupposes not merely a being who is in the world but, more than that, one whose relation to the world includes a basic competence with the milieu of every encounter. And we might anticipate that without some such competence the notion of surprise would be meaningless not because it could never occur, but to the contrary because it would define every experience in which a datum is given to a consciousness that would be, by definition, wholly unprepared for it.

This is already enough for us to recognize that even when it is a matter of a datum that arrives from beyond expectation, *the intelligibility the experience* depends in some important measure on the condition of the subject and its relation to the world in which it lives. I therefore situate my own thoughts within a general hermeneutics of surprise, by which we may expect to differentiate among forms and degrees of surprise by attending to the context in which they occur (historical, cultural, personal-psychological, etc.), though without surrendering the possibility of grasping their inner unity in some irreducible essence (*eidos*). Inevitably, much of this work will proceed by way of specific examples. On this occasion, my example will be the episode commonly called 'the conversion of St. Paul.'

2 Paul on the Road to Damascus

Few to none of us are likely to have undergone greater surprise than did Paul on the road to Damascus, and few with greater consequences. An entire way of life was at stake, and indeed the relationship between entire kindred, though distinct, traditions. Let us recall the account of Paul's experience given in the *Acts of the Apostles*:

> But Saul, still breathing threats and murder against the disciples of the Lord, went to the high priest and asked him for letters to the synagogues at Damascus, so that if he found any belonging to the Way, men or women, he might bring them bound to Jerusalem. Now as he journeyed her approached Damascus, and suddenly a light from heaven flashed about him [*exaiphnes te auton periestrapsen phos ek tou ouranou*]. And he fell to the ground and heard a voice saying to him, "Saul, Saul, why do you persecute me?" And he replied, "Who are you, Lord?" And he said. "I am Jesus, whom you are persecuting: but rise and enter the city and you will be told what you are to do." The men who were traveling with him stood

[3] See especially P. Ricoeur, *Oneself as Another* (Chicago: University of Chicago Press, 1992), pp. 335–341 and *The Course of Recognition* (Cambridge, MA: Harvard University Press, 2005), pp. 154 and 261–262. In these passages, Ricoeur resists any attempt to absolutize either the subject or the other, and develops a conception of (mutual) recognition that would oppose any general account of unqualified surprise.

> speechless, hearing the voice but seeing no one. Saul arose from the ground; and when his eyes were opened, he could see nothing; so they led him by the hand and brought him into Damascus. And for three days he was without sight and neither ate nor drank.[4] Acts 9:1–9

In his few references to this event, Paul seconds some of the key features in the Lucan account (in discussing these events, I will refer uniformly to "Paul," though the man named Saul does not take that name until some time after the experience on the road to Damascus). Defending the authority of his message, he tells them "I did not receive it from man, nor was I taught it, but it came through a revelation of Jesus Christ [*apokalypseos Iesou Christou*]" (Gal 1:12). And to the Corinthians he claims that although he came after the apostles who lived and walked with Jesus, nonetheless "he appeared also to me [*opthe kamoi*]" (1 Cor 15:8) and I have seen Him [rhetorically: *Iesoun ton kyrion hemon heoraka*?] (1 Cor 9:1). Let us note first that whatever "suddenly" happened to Paul was a matter of revealing, seeing, and appearing. Before exploring this nexus of themes further, we should immediately consider two points about the seeing that Paul invokes in 1 Corinthians 9. On the one hand, his word *heroaka*, which derives from *horao*, has only a general sense of perceiving or even simply experiencing that does not, by itself, fully support an association specifically with ocular vision. One thinks rather of seeing with the mind or seeing spiritually. On the other hand, there can be no question that a strong sense of the reality of what is seen—that it truly was Jesus—was crucial for the authority of the apostolic teaching (see Acts 1:3), including, as I have already noted, Paul himself. This is also the sense of his word *apokalypseos*, which signals knowledge achieved in the coming to light of what was previously unseen. At the heart of the event in the road to Damascus is a revelation by which it suddenly becomes possible for Paul to know something previously invisible to him so long as he was seeing according to the powers and from the perspective that had served him to that point.

This brings us to a second factor, at least if we turn return to Acts 9. Those with whom Paul traveled did hear a voice, but saw no one. Somewhat later, Paul himself tells the Galatians and Corinthians that Jesus was revealed and appeared to him. Here we have a puzzle only so long as we continue to equate "seeing" with "ocular vision." The puzzle disappears when we envision everyone present to have seen a "light from heaven" and heard a voice, but only Paul to have understood—if we envision only Paul to have seen Jesus in the sense of knowing what His words meant. For Paul's companions, whatever the voice said will have been unintelligible, or perhaps intelligible but untrue—for if they not only heard a voice but also grasped its words, then we must suppose that they immediately discredited them. After all, they were not swayed by them so much as perplexed and quite possibly frightened. As for Paul, we should be careful in our interpretation of his response to what had suddenly befallen him. If in contrast with his companions he plainly was moved in some way by what he had heard, still there is no reason to necessarily

[4]All New Testament passages are taken from E. Nestle and K. Aland's *Greek-English New Testament*, which follows the Revised Standard Version of the English translation (Stuttgart: Deutsche Bibelgesellschaft, 1990).

conclude that he accepted everything then and there. Acts 9 tells us only that he was struck blind, and in his letter to the Galatians he seems to indicate that he spent another 3 years in "Arabia" before making his way to Jerusalem and then on to his ministry to the Gentiles (Gal 1:17–18). But even here, we should move slowly. Did Paul begin immediately to proselytize there, to the uncultured Nabateans, only to return without leaving any record of success (and it is very difficult to believe that he would have kept any successes to himself)? Or were the years in Arabia spent digesting the impact and the import of his surprise on the road to Damascus, whereupon he could meet James and Peter in full possession of his new identity? Lacking a means to decide the question, we can at least say that the light and the voice left an impression already then and there on the road that made it impossible for him to continue as he was. And if at first this meant only that he was unable to proceed with plans to obtain letters by which to identify and persecute more followers of Jesus, there is nonetheless general consistency between that turn and what eventually becomes a mission to preach the resurrection of Jesus Christ. If the fact that Paul, unlike his companions, was stunned by what he saw and heard requires us to think that it reaches him in a way that it did not reach them, the fact that his subsequent path moves progressively away from his hostility to the followers of Jesus suggests that what thus reached him had a specific content that made an irrecusable claim on his understanding even of himself.

Now we should not fail to recognize that such a claim—powerful and yet intimate—is certain to have an effect on the body of the one who is claimed.[5] Here even more than elsewhere, there is little by which to find our way. As I have already noted, Acts 9 tells us that the experience left Paul blind for three days. Acts 22 has Paul repeat essentially the same story as Acts 9, including the fact that he was temporarily blinded. Among scholars of Paul there is considerable speculation about the medical nature of his condition, but the most widely accepted view seems to be that he suffered from an affliction that rendered him prone to infection and inflammation of the eyes. And this does square well with certain passages that might otherwise elude our understanding: at one point, Paul refers to a "body ailment" that rendered him disgusting to those who saw him (Gal 4:13–15), and at another he mentions a "thorn in the flesh" received from God in order to prevent him from pride at his apocalyptic vision (2 Cor 12:7–10). Whether, as the author of *Acts* and Paul himself clearly think, his temporary blindness was caused by an apocalypse of Jesus the Christ, or only provoked by what he thought he saw and heard on the road to Damascus, it is evident that that blindness itself is a corporeal expression of Paul's

[5]The Pauline conception of the body deserves extended attention. As Nietzsche noticed, in Paul's distinction of *sarx* (flesh) from *soma* (lived body), there is the seed of a pluralization of wills. Paul's call that we cleave to the resurrected body of Jesus is thus a call to unification of an inchoate willing that would also be a transfiguration of our corporeality. Yet this thought would have to be reconciled with the violent corporeal dissociation that seems to occur in his mystical experiences. See, for example, 2 Cor 12: 2–3: "I know a man in Christ who fourteen years ago was caught up into the third heaven – *whether in the body or out of the body I do not know, God knows*. And I know that this man was caught up into Paradise – *whether in the body or out of the body I do not know, God knows…*" (emphasis added).

profound sense of surprise. We may leave it to the psychoanalysts to interpret the possible significance of a blindness that ensues from a light in which Paul saw what he would soon enough describe as Christ's victory over death. Let us instead hold fast to a singular fact: this victory will not have been achieved from within the Jewish tradition in which Paul had been raised and schooled, and of which he had been a fervent advocate, but by someone whose followers he had considered deserving of the most aggressive persecution. This thought, at the very least as a real possibility, will have thrown Paul into the grave crisis that surely lies at the root of his physical complaint.

And so at last we come to the question of Paul's emotional state when he heard and saw Jesus. As we have seen, this occurred in a revelation that came not at his own initiative and such that he alone received it in its fullness. Paul was thus singled out, set apart from his companions as well as, of course, the comfort of his own identity—though without necessarily passing immediately into the Christian faith that defined his later public life. This way of seeing the episode leads us to consider, as a working hypothesis, that such a call must have precipitated anxiety. We may also consider, as a further specification of the hypothesis, Heidegger's notion that anxiety is a matter of attunement to one's own groundlessness, or to what he has called the "abyss of meaning."[6] Now for this interpretation of Paul to hold, we would have to think that between his Jewish and Christian faith there was a moment—perhaps a long moment—in which he not only was without any faith at all but was left feeling that any meaning of any kind in fact rests on nothing. Let me simply note, for the moment, that this feeling would already be an interpretation, at the level of affect, of what had reached him so unexpectedly on the road to Damascus. Of course, we might parse the experience further, and ask whether there is a difference between the event of the call itself, the moment in which Paul was surprised by it, and the anxiety that may have ensued. Such a proposal does help us to see more clearly that Paul was properly surprised only upon assigning to the call some meaning that pertained directly to him in his worldview, at which point there arose in him an anxiety that further interprets the meaning of the surprise (i.e., as impossible to situate in line with his previous beliefs). In short, unless the surprise does not already entail within itself some defined meaning, the rise of any particular emotion would be unintelligible.

But even with these refinements in mind, we should not conclude too easily that what Paul felt upon hearing the voice of Jesus was specifically anxiety, at least as Heidegger has helped us to understand it. After all, this would leave unanswered the question of just how he regained faith in the one God after he would by definition has have lost it. Moreover, precisely what anxiety seems necessarily to suspend—openness to divine transcendence—will have been a feature of Paul's subjectivity both first as a Jew and then subsequently as a Christian. Would this faith be only continuation across a hiatus, or is it the unifying principle of an entire way of being that is, to be sure, compelled to redefine itself? There are two ways to approach this question. We may attend to what we know of Paul's Judaism as he set out on the

[6] M. Heidegger, *Being and Time* (London: SCM Press, 1962), p. 194.

road to Damascus, and we may attend to the Christianity that is abundantly available in his own letters. The former approach seeks an account of Paul's *disposition* as Jesus suddenly called to him. The latter may provide us with a sense of how he *integrated* the experience.

3 Paul as Jew

We know that at least by the time of young adulthood Paul was an adherent to Pharisaic Judaism, a movement whose zeal was expressed in attempts to memorize much of the Hebrew bible and a fervent commitment to the law as defined by a personal master, or "father." When at about age 21 Paul made his way from Tarsus to Jerusalem, it was in order to advance in his learning under such a master. This he did for 4 years, emerging finally as a fledgling master of Pharisaism in his own right. The name of his own master during his training is not known, but we do know the name of a great teacher who, by Paul's own account, affected him greatly. This was Gamaliel, who was born into the third generation of a family of prominent leaders also esteemed as religious authorities. The author of *Acts* takes note of this in chapter 34, referring to Gamaliel as a "Pharisee" and "a teacher of the law, held in honor by the people." His influence on his people seems to have been immense, extending beyond the impact of his interpretation and formal judgments concerning the law to include an instruction by example that could involve virtually anything that he was seen to do. In this way, Gamaliel touched all of the Pharisaic community if not directly, then according to the measure in which his important presence had to be taken into the reckoning of any teaching proposed by lesser masters. This is probably how to understand the statement attributed to Paul in Acts 22:3, to the effect that he studied in Jerusalem "at the feet of Gamaliel."[7]

Now of Gamaliel's views, the traditions have recorded only a few statements, though in the context of this investigation at least two of them are certainly of interest. In the *Mishnah*, he is said to have taught "Secure a teacher for yourself, [in religious questions] remove yourself from doubt the doubtful, and do not tithe in excess by general estimation."[8] The entirety of this teaching is in recognizable solidarity with the main lines of Pharisaism, but the emphasis on the importance of a teacher must also be understood in context. As I have already noted, Gamaliel himself was a teacher of extraordinary influence. In fact, he is often thought to be the first in the Jewish tradition to have been honored with the title of "Rabban." Although a commitment to authoritative religious teaching was hardly new in the time of

[7] This conception is consistent with doubt expressed by Helmut Koester that Paul could literally have been a student of Gamaliel, since the same sentence that seems to make such a claim also makes the obviously false claim that Paul grew up in Jerusalem ("...brought up in this city at the feet of Gamaliel"). See H. Koester, *Introduction to the New Testament, vol. 2. History and Literature of Early Christianity* (Berlin: Waltr de Gruyter, 1982), p. 107.

[8] *Pirkei Avot* 1:16.

Gamaliel, previously it was identified more strictly with judgeship, as distinct from the roles of kingship, prophecy and ritual priesthood. But in Gamaliel's own time, the roles of kingship and ritual priesthood were already in decline, and the age of the prophets was widely thought to have ended, and so his exhortation to "secure a teacher" must have had a particular appeal to his people--among them, plainly enough, Paul.

A second of Gamaliel's teachings is found in an intervention recorded by the author of *Acts*: learning that a group of zealous Pharisees wished to put to death some of the apostles, Gamaliel reminded them of similar, regrettable events in recent times and then advised, "So in the present case I tell you, keep away from these men and let them alone; for if [the apostles'] plan or this undertaking is of men, it will fail; but if it is of God, you will not be able to overthrow them. You might even be found opposing God!" (Acts 5:38–39). Of course, lacking confirmation among Jewish sources, the authenticity of this citation is open to doubt, yet it displays a tolerance that is consistent with the Hillel school of which Gamaliel was the head.

These two passages support the following thought about Paul the Pharisaic Jew: he not only, like many other Jews of the day, sought a great teacher, but knew that the one who enjoined such a search also was prepared to tolerate the fledgling movement following Jesus Christ. That said, we can only suppose that this was a notion familiar to Paul, and evidently not an accepted rule for his behavior. Not long after Gamaliel is said to have urged tolerance, Paul was found at the stoning of Stephen (Acts 8:1).

The difficulty is not unrelated to another conflict in the Judaism of Paul's day that will have contributed to his disposition as he departed Jerusalem for Damascus. On the one hand were the Pharisees, himself surely included, who were intent on developing an oral tradition that would protect the Law, so that

the marks of Jewish fidelity and election could remain clear. On the other hand was a marginal apocalyptic movement in which, without questioning the importance of the Law, was intent on the end of all ages defined by a messianism understood more as a transcendental condition of the relation with God than as the achievement of a single historical figure. Here again, it is dangerous to suppose more than a notion, or perhaps a possibility, known to Paul as he executed what he continued to understand as the proper, Pharisaic definition of Jewish life.

Still, these two simple hypotheses concerning what Paul knew and perhaps had in mind would go a long way toward better understanding some prominent features of the experience on the road to Damascus. Let us recall, first, that it was Paul alone who not only heard a voice, but also "saw" that it was Jesus. Perhaps his complex relation to his own Jewish faith included elements that rendered him disposed to somewhat more than immediate rejection of the words spoken to him by Jesus. And so, second, let us also keep in mind the fact that what was thus revealed to him, regardless of how much time we suppose it took him to fully understand it, affected the Jewish dimension of his identity much more significantly than it did the Greek and Roman dimensions. After all, Paul never entirely gave up the habit of invoking his Roman citizenship when it suited him (e.g., Acts 22:25), and his Greek education—such as it was—was never an important factor in the account he gave of himself to others. His Greekness and Romanness, in short, remained in loose complementarity with the rest of himself, but his Jewishness henceforth was defined specifically as the antecedent of his fidelity to Jesus Christ.

What, in view of these things, are we to make of the surprising event itself? Was there anything in Paul's Jewishness—in the Judaism of the time, such as would have been known to him—that may have prepared him to take serious the voice of Jesus as it called to him by name, rather than dismissing it out of hand? As it happens, there may have been in Paul, and in many Jews of that time, a particular disposition to hear the voice of God. Of course, claims for such an experience appear frequently in the Jewish bible. One thinks most readily of the "still small voice" heard by Elijah (1 Kings 19:12–13), as well as the voice of God ringing from on high to Ezekiel (Ezek 1:25) or from Zion to Amos (Amos 1:2). But these were experiences of the Prophets, and as I have already noted by the time of Paul it was thought that the age of the prophets had ended. And so, as if to answer to people's desire for divine manifestation, there emerged, within two centuries of Paul's own time a new conception of the divine self-communication in a voice that may on occasion reach any fervent believer—and no longer only those who had the gift of prophecy. As one might expect, the notion was controversial. Some teachers saw in it the sign of a new age and thus a new development in the relation to God. Others feared a decline into superstition and urged the supremacy of Law. In any case, the name for this divine voice is *bat-ḳol*, which, interestingly enough, signifies something more like echo of the original voice rather than the voice in its pure form. For God's own voice was thought to be supersensible, whereas in order for it to reach a person lacking the special gift of prophecy, it must become audible. To this should be added only the reminder that within the broader Jewish tradition of divine communication by voice and call, there are instances in which the one who *hears* the voice does also *see* something, though not God as God. Moses famously saw only God's back as he passed him in the cleft atop Mt. Sinai (Ex 33:23) and in Ezekiel's case the voice of God is either accompanied by "the hand of the Lord" or delivered angels who are in some features anthropomorphic. Let us then suppose, again without overworking the point,[9] that while there is no evidence at all that Paul actively adhered to this notion, it is more than possible that he would have known of it—and indeed, given the unmistakable fervor of his own faith during a time of some trial for the Jews of Jerusalem, might well have at least sympathized with its evident motivation. And so if brief exploration Paul's relationship to Gamaliel and some cautious speculation about his possible awareness of an apocalyptic strand of first century Judaism have led me to propose that Paul's experience of surprise on the road to Damascus may have been prepared by elements that disposed him to understand it in a specific way,

[9] As a matter of fact, not everything in the notion of *bat-ḳol* fits easily with the thesis that it would have been an element of Paul's disposition to hear the voice of Jesus. For example, voice is often thought to be female. For a helpful overview of this tradition and some important literature, see the entry for *bat-ḳol* in the online *Jewish Encyclopedia*: http://www.jewishencyclopedia.com/articles/14725-voice-of-heaven

I will now add that we might begin to see the basis for a desire to hear and see God in a context where Paul and his community had thought, no doubt to their great distress, that God had fallen silent.

If these many speculations hold up, then we might conclude of Paul that he was not so much *converted* from one way of life to another that is entirely different, as he was *called* by a voice that spoke to a longing in his heart that had hitherto been directed by a range of commitments that turned *him from away from* Christian faith, the very sight of which incited in him an all too ready aggression—but without those commitments closing him from the possibility of a positive reckoning. To be sure, Paul did change his manner and his outlook to a degree that merits the word conversion, but there is no longer any reason, if there ever was one, to think that this resulted solely from the suddenness of a divine initiative. To the contrary, properly speaking the conversion took place within Paul, as he registered the call of Jesus within the rich array of notions and principles available to him, and realized himself to be someone other than he had previously thought. Whether or not this precipitated anxiety such as Heidegger understands it seems doubtful, since nothing supports the thought that Paul ever lost all faith that God exists. What now seems much more likely is that his great surprise left him troubled as to who God truly is and what might be his own relation to God. And this of course is the very substance of his preaching. Let me now turn, much more briefly, to the Paul whose new faith leads him so far as to declare "it is no longer I who lived, but Christ who lives in me." (Gal 2:20)

4 Paul as Christian and as Mystic

If I turn now to Paul's own perspective on the voice he heard on his way to Damascus, this is in part because it seems impossible to say more about the experience starting from an antecedent Jewishness about which we in fact know very little. There are two difficulties that must be faced by any attempt to interpret Paul's experience starting instead from his Christian orientation. I have already touched on the first when attending to his few explicit references to the matter: not only are they few in number, but they also tend always to concentrate on establishing a claim to religious authority. One is thus led to ask somewhat more generally whether the person that Paul became as a follower of Christ might express something about the event by which it became possible. It is here that one meets a second difficulty: we cannot rule out the possibility that in his writing, which is to say his *preaching*, Paul adapted his account of himself to the demands of his task.

With all due caution, I restrict myself to the following matters:

1. Paul's writing is without exception aimed at preaching the gospel (*euanggelion*)[10] of Jesus Christ, by which is meant, writ large, that the Resurrection of Jesus

[10] As is often noted, *euanggelion*, like the Aramaic *besora* (Paul would have known both), denotes not only news that is not only "good," but even triumphant. See, e.g., B. Chilton, *Rabbi Paul* (New York: Doubleday, 2004), pp. 61–62.

Christ fulfills the messianic hope long preserved in the Jewish tradition in which he had been formed.
2. This preaching is directed not only to the Jewish people, but all humanity, whether Jew or Greek, slave or free, male or female (cf. Gal 3:28).
3. This universalization of the Christian message is rooted, in Paul's case, in a distinctive conception of the human person—or, if one prefers, of our *subjectivity*—that is derived from Paul's own understanding of what was revealed to him on the road to Damascus.

Earlier, I observed that Paul's preaching emanates from an assimilation of his life with that of Jesus Christ. The theme of Christ in us or us in Christ (Paul appears untroubled by the possible difference) appears throughout his correspondence (e.g., Rom 8:10, 2 Cor 13:5, Gal 4:19, etc.), doubling as a description of our true condition and as an admonition for how to live. There is no mistaking the point of arrival for Paul, as he integrated the meaning of the words he heard on the road: overwhelmed by them, feeling himself and his convictions put suddenly in question, Paul submitted to the message entirely, taking it in as the new principle of his life and understanding. Identifying with one who had willingly given himself up to crucifixion, Paul, as it were, followed suit *in extremis*, emptying himself of all that centered on his own will in favor of all that is commanded by the will of Christ. Philosophically, this is recognizably an instance of decenterment. Theologically, its specific form is Christological. And let us not forget that the Pauline message extends to all human beings. The Pauline subject is decentered by absolute commitment to Christ, wherein he finds the truth of our condition.[11] In short, the surprise on the road to Damascus proves to have a mystical dimension in which, however, is grasped a universal proposition concerning humanity.

This outcome leaves us with an open question that is, I think, of some interest for any hermeneutics of surprise and understanding. What we see in the full trajectory of Paul's experience is a passage from an experience that is, in its concrete form, essentially *inward* and *personal*, to a discourse that aspires to become, in its *outward* expression, fully *universal*. With regard to Pauline thinking as such, we have in view what might thus be called a "democratization of mysticism."[12] In its proper context, the formula signals Paul's subversion of spiritual elitism: an experience that is often thought to be reserved for the select few yields claims about a way of life available to all. Of course, this is not entirely unexpected for a discourse that is deliberately evangelical. But the *philosopher* must not fail to ask about the possibility of such a translation. For unless there is an affinity—psychological, cultural, linguistic, and perhaps all of these—between that which surprises and that which is

[11] This Christocentric view opposes the recent attempts of G. Agamben and A. Badiou, for whom the decenterment of the Pauline subject is aligned with an allegedly anarchic relation to the Law. It is not that the new Paul henceforth prescinds from consideration of the Law, but that when he hears Christ he understands that the Law is fulfilled by an abundance of love.

[12] U. Luz, "Paulus als Charismatiker und Mystiker, in T. Holz, *Exegetische und theologische Studien. Gesammelte Aufsaetze II* (Leipzig: Evangelische Verlaganstalte, 2010), p. 89. Quoted in D. Marguerat, "Paul le mystique," in *Revue theologique de Louvain* (43) 2012, 4: 489.

interpreted as the surprise, the event itself is literally unintelligible. The various stands of what I have just called 'affinity' belong to what ordinary language calls 'a context.' To be sure, the peculiarity of a surprise lies in its capacity to alter the entirety of the context, bathing it in a new light or perhaps a new color. We are therefore tempted to suppose that it comes to us, as it were, with its own horizon of meaning. Yet unless this at least some of this new meaning can be fused with the meaning of what it may challenge and transform, it remains strictly alien. The nature and limits of that fusion are open to interpretation and call for concepts that do not obscure the experience in question.

Surprise, Meaning and Emotion

Claudia Serban

Abstract The paper examines the narrative construction of *Alice's Adventures in Wonderland* (1865), undisputed masterpiece of nonsensical literature, and shows that it is moved by an avalanche of surprises which obey to a complex mechanism in which cognition and emotion constantly interact. In the light of nonsense, not only will the intertwining of the cognitive and the emotional appear differently, but also the nature of the surprise caused by the lack of meaning will be particular. Rather than the unexpected (or the marvelous) in regard to cognitive anticipations, surprise will then designate the unexpected fulfillment or deception of desire. Furthermore, when the playfield of surprise is no longer perception or experience (in a very broad sense), but language, we will be confronted to the dark face of cognitive emotions: that of incomprehension. Therefore, the eruption of surprise in language, if it occurs *outside* the logic, will still have an effect *on* logic: that of assigning its limits. And while the *marvelous* becomes *absurd*, the increasing amount of incomprehension will favor the development of new, unexpected emotions, such as *anger*, *disappointment* and *indignation*. The abundant presence of such negative emotions confirms that, in *Alice*, the experience of nonsense is at least as important as that of the marvelous. It is thus the double logic of sense and desire that provides the measure of surprise in the Carrollian Wonderland, explaining the variety of emotional responses to (un)satisfied desire and to the lack of meaning.

Keywords Surprise · Disappointment · Unexpected fulfillment · Carroll · Marvelous · Language · Absurd · Incomprehension

The impossibility of dissociating knowledge and affectivity otherwise than artificially legitimates the fact that we speak of cognitive emotions to designate, for instance, not only the joy of intellectual or scientific discovery, but also the cognitive motor of (philosophical) astonishment. Both the reaching of a much-attended

C. Serban (✉)
Université Toulouse – Jean Jaurès, Toulouse, France

N. Depraz, A. J. Steinbock (eds.), *Surprise: An Emotion?*, Contributions To Phenomenology 97, https://doi.org/10.1007/978-3-319-98657-9_9

scientific result and the primary astonishment that motivated the research behind it can be considered, furthermore, as patterns of surprise, the former in the sense of the sudden and pleasant replenishment of an intellectual appetite, the latter as a striking and frustrating inability to understand that will motivate the quest for knowledge. What happens instead when the intertwining of the cognitive and the emotional appears in the light of a constant exposition to non-sense or to the lack of meaning? What kind of surprise is then at stake – if one can still speak here of surprise?

It is true that if we generally designate by the term "surprise" an unexpected event, the unexpected here at stake does not necessarily involve specific cognitive expectations: a surprise-party or an unexpected meeting will not contradict my cognitive convictions. The matter is somewhat different when the surprise is generated by what someone says, that is, when the playing field of surprise is no longer perception or experience (in a very broad sense) but language. But here as well, a difference is to be taken into consideration: a surprising statement (for instance, "I definitely prefer rainy days to sunny days.") can simply contradict my momentary convictions or simply common sense; but it can also contradict, more generally, the tacit rules of meaning, in the case of *true* non-sense (if someone told me instead: "I definitely prefer rainy plums to sunny melons"). The question that I would ask then, expressing my surprise, is simply: what do you *mean*? The answer here could be: "plums have the taste of rainy autumns and melons that of sunny summers." In this case, the coherence of meaning would be restored by the deciphering of the metaphor, even if the message remains primarily counter-intuitive and thus surprising. But what if the lack of meaning grows greater and greater, becoming more and more pregnant? What kind of surprise would I then experience, and how would I emotionally react to it?

When the unexpected occurrence that provokes my surprise is not a novelty that troubles or colors my everyday life but a message that undermines the conditions of meaning and comprehension (the formal and material coherence of speech), I will be puzzled rather than amazed, which is to say that I will experience the dark face of cognitive emotions: that of incomprehension, of the negative gain of knowledge that is also lack of satisfaction. My surprise will be mixed with perplexity when confronted to the meaninglessness of speech, and I will aim to find a meaning to what seems at first sight meaningless (such as speaking of rainy plums and sunny melons).

In his *Tractatus Logico-Philosophicus*, Wittgenstein states that "there are no surprises in logic,"[1] but not every speech act is under the rule of logic, and the eruption of surprise in language, if it occurs *outside* logic, can still have an effect *on* logic – at least that of assigning its limits. This is the case with what we can call, following Natalie Depraz, "the surprise of non-sense."[2] To our previous question (i.e. How will the intertwining of the cognitive and the emotional appear in the light of non-

[1] See Ludwig Wittgenstein, *Tractaus Logico-Philosophicus*, trans. C.K. Ogden (Mineola, NY: Dover, 1999), 6.1251. "*Darum kann es in der Logik nie Überraschungen geben.*"

[2] Natalie Depraz, "The Surprise of Non-Sense," in *Enactive Cognition at the Edge of Sense Making*, eds. Massimiliano Cappuccio and Tom Froese (London: Palgrave MaMillan, 2014), 207–237.

sense?) we can add indeed the more specific question of the particular nature of the surprise and the emotion caused by the lack of meaning. In order to analyze it, we will follow here a somewhat unusual path, by examining the narrative construction of *Alice's Adventures in Wonderland* (1865), an undisputed masterpiece of nonsensical literature. Its author, Lewis Carroll (Charles Lutwidge Dodgson) was an Oxford mathematician and logician, who might have agreed to consider his literary exercises as a form of what Wittgenstein later would call "language games." Nonsense literature can indeed be regarded as a language game that abundantly cultivates the surprises that logic cannot host, that is, the surprises that represent a challenge to meaning, thus signing the autonomy of fiction in respect to logic and assigning its limits to logic. If we focus on the narrative construction of *Alice's Adventures in Wonderland*, it will appear to us as if it proceeded by series of avalanche-like surprises that obey, as we shall see, a complex mechanism in which cognition and emotion constantly interact and nourish one another.

Among the philosophical readings of Lewis Carroll's *Alice*, the one provided by Gilles Deleuze in *The Logic of Sense* not only pays great attention to the paradoxes of sense but also considers these very same paradoxes as "pure events."[3] It is our intention here to show that these events have a more specific name, that of *surprises* of a very particular type. The whole correlate of the surprise-event is Alice's *adventure*, namely, *someone's* lived experience, in which the surprise of non-sense is the trigger of existential interrogations[4] and of a process of becoming aware. The deconstruction of meaning under the pressure of non-sense leads ultimately to a crisis of personal identity, inviting Alice to reconsider herself in the light of her (surprising) adventure(s). The logic that the eruption of nonsense challenges and questions is consequently anything but an impersonal logic; quite the contrary, the reference to the *subject* (furthermore, to a very profound subject: the subject of desire) is constitutive for it. In Lewis Carroll's *Alice*, the intertwining of the cognitive and of the emotional is given indeed within the constant interaction between a *logic of sense* and a *logic of desire*: the emergency of surprise is simultaneously measured against the requirements of meaning and against the requirements of desire. The Carrollian *Wonderland* is simultaneously, according to the double meaning of the verb: *to wonder* – to be marveled and to ask oneself – a land of marvel and a land of interrogation,[5] a territory in which the challenge of non-sense is launched from the profound maze of desire. Therefore, rather than the unexpected (or the marvelous), in regard to cognitive anticipations (a somewhat poor dimension in *Alice*, given that the protagonist is a young child), surprise here designates the unexpected fulfillment or deception of desire. For this reason, the valences of surprises are diverse, and so are its connotations that can go from astonishment to fear and anger.

---

[3] Gilles Deleuze, *Logique du sens*, (Paris: Minuit, 1969), 6; *The Logic of Sense*, trans. Mark Lester with Charles Stivale (New York: Columbia University Press, 1990), 2.

[4] See Sophie Marret, *Lewis Carroll. De l'autre côté de la logique*, (Rennes: Presses Universitaires de Rennes, 1995), 83.

[5] As highlighted by Sophie Marret, *Lewis Carroll*, 218.

In order to test all these hypotheses on Carroll's nonsensical fiction, we shall start our direct confrontation with the text and "begin at the beginning":

> Alice was beginning to get very *tired* of sitting by her sister on the bank, and of *having nothing to do*. Once or twice she had peeped into the book her sister was reading, but it had no pictures or conversations in it, "and what is the use of a book," thought Alice, "without pictures or conversations?"
>
> So she was considering in her own mind (as well as she could, for the day made her feel very sleepy and stupid), whether the pleasure of making a daisy-chain would be worth the trouble of getting up and picking the daisies, when *suddenly* a White Rabbit with pink eyes ran close by her.
>
> There was *nothing so very remarkable* in that, nor did Alice think it so very much out of the way to hear the Rabbit say to itself, "Oh dear! Oh dear! I shall be too late!" But when the Rabbit actually *took a watch out of its waistcoat-pocket* and looked at it and then hurried on, Alice started to her feet, for it flashed across her mind that *she had never before seen* a rabbit with either a waistcoat-pocket, or a watch to take out of it, and, *burning with curiosity*, she ran across the field after it and was just in time to see it pop down a large rabbit-hole, under the hedge. In another moment, down went Alice after it, *never once considering how in the world she was to get out again*.[6]

At the beginning of the first chapter, we find Alice sunk in a state of boredom and inactivity that does not allow particular expectations and that is *suddenly* interrupted by the apparition of the White Rabbit. While definitely unexpected, this apparition is not yet a genuine source of surprise: it takes a talking Rabbit that carries a watch in his waistcoat-pocket for such an effect. And the surprise is lived here as *burning curiosity* – a figure of the intertwining of the cognitive and the emotional that marks the beginning of the adventure by pushing Alice to leap down the rabbit-hole, decision that requires leaving aside any expectation and anticipation.

But the fall through the rabbit-hole is long, creating an expectation- and interrogation-leap: "Either the well was very deep, or she fell very slowly, for she had plenty of time as she went down to look about her and *to wonder what was going to happen next*."[7] The fact that expectation is here affective rather than cognitive, related to desire rather than to a thirst for knowledge, is shown by the episode of the marmalade jar: Alice "took down a jar from one of the shelves as she passed; it was labelled 'ORANGE MARMALADE,' but to her great *disappointment* it was empty."[8] Alice's first specific expectation thus resolves itself negatively into *disappointment*. But this doesn't suffice to stop the interrogative movement that the fall through the rabbit-hole triggered: "Down, down, down. Would the fall never come to an end! [….] I *wonder* what Latitude or Longitude I've got to? (Alice had no idea what Latitude was, or Longitude either, but thought they were nice grand words to

[6] Lewis Carroll, *Alice's Adventures in Wonderland*, in *The Philosopher's Alice*: *Alice's Adventures in Wonderland and Through the Looking-Glass*, ed. Peter Heath (London: Academy Editions, 1974), 13–14. In order to avoid the constant notation "added emphasis," throughout this essay (as in this quote) we will italicize the terms that mark the temporal, affective and cognitive dimension of surprise (or of anticipation, deception, desire…)

[7] Carroll, *Alice*, 14.

[8] Carroll, *Alice*, 14.

say (added emphasis)."[9] Meaning, as we can see, is not truly a requirement for speech here; furthermore, Alice progressively sinks into a sort of semantic reverie governed by meaninglessness: "'Do cats eat bats? Do cats eat bats?' and sometimes, 'Do bats eat cats?' for, you see, as she couldn't answer either question, it didn't much matter which way she put it."[10]

But the fall ends as unexpectedly as it began ("when *suddenly*, thump! thump! down she came upon a heap of sticks and dry leaves, and the fall was over"[11]), thus multiplying the interrogations concerning what might happen next. The sudden discovery of a small key provides Alice with the hope that she will be able to get out of the small place wherein she is enclosed. But here comes a new disappointment, for the small key doesn't fit in any of the doors around it. This new deception contributes nevertheless to an increasingly strong and yet totally irrational desire: "'Oh, how *I wish* I could shut up like a telescope! I think I could, if I only knew how to begin.' For, you see, so many out-of-the-way things had happened lately, that *Alice had begun to think that very few things indeed were really impossible*."[12] Against all expectations, the irrational desire will nevertheless be fulfilled when Alice drinks from the small bottle, and rather than disappointment there arises the satisfaction associated with a fulfilled desire.

But this satisfaction is cruelly momentary: If she is now sufficiently small to pass through the small door, Alice has no longer access to the key on the table. Desire is now powerless and Alice has no choice but to accept what will happen next – that is, to truly assume her adventure. This new receptivity, however, is combined with an expectation of the marvelous and a rejection of fad and boring familiarity: "Alice had got so much into the way of *expecting nothing but out-of-the-way things to happen*, that it seemed quite dull and stupid for life to go on in the common way."[13] An expectation that nevertheless does not *predict* anything and yet will be fulfilled, provoking the first genuine (and textually attested) experience of surprise in the narration (at the beginning of chapter II): "'Curiouser and curiouser!' cried Alice (she was so much *surprised*, that for the moment she quite forgot how to speak good English); 'now I'm opening out like the largest telescope that ever was!'"[14] The first (literal) eruption of surprise in *Alice* significantly manifests itself as a *disturbance in language*, suggesting that, when the logic of experience is compromised, language cannot remain the same: surprise and nonsense are here intimately bound. This joint apparition causes a new wave of interrogations:

> 'Dear, dear! How queer everything is to-day! And yesterday things went on just as usual. *I wonder* if I've been changed in the night? Let me think: was I the same when I got up this morning? I almost think I can remember feeling a little different. But if I'm not the same, the next question is, Who in the world am I? Ah, that's the great puzzle!'[15]

---

[9] Carroll, *Alice*, 15.

[10] Carroll, *Alice*, 16.

[11] Carroll, *Alice*, 16.

[12] Carroll, *Alice*, 17–18.

[13] Carroll, *Alice*, 20.

[14] Carroll, *Alice*, 21.

[15] Carroll, *Alice*, 24.

The interrogation here concerns nothing less than personal identity, whose questioning is the ultimate effect of the immersion in Wonderland and of the exposure to nonsense and surprise.

The interrogative climate is furthermore propitious to new emergences of surprise – even in the particular figure that one could call *self-surprise*: "As she said this she looked down at her hands, and was *surprised* to see that she had put on one of the Rabbit's little white kid gloves while she was talking. 'How can I have done that?' she thought."[16] To no longer know who *I* am implies that *I* become able to surprise myself. This is how we could resume the experience that Alice has here. The unpredictable mixture of expectations, events, and deceptions that Alice's adventures in Wonderland contain lead her to be astonished by her own conduct but also to nonchalantly attempt the impossible:

> 'Would it be of any use, now,' thought Alice, 'to speak to this mouse? Everything is so out-of-the-way down here, that I should think very likely it can talk: at any rate, there's no harm in trying.' [...] 'Perhaps it doesn't understand English,' thought Alice; 'I daresay it's a French mouse, come over with William the Conqueror' [...] So she began again: 'Où est ma chatte?' which was the first sentence in her French lesson-book. The Mouse gave a sudden leap out of the water, and seemed to quiver all over with fright.[17]

Between humans and animals, the barrier of communication could be, not language in general, but a particular language rather than another. This is Alice's hypothesis while addressing the mouse in French, and this will allow her to get, if not an answer, at least a reaction. The improbable expectation thus realizes itself and desire arrives to satisfaction.

Yet language soon reveals itself as a field of misunderstanding, and the increasing amount of incomprehension in which the *marvelous* becomes *absurd* will favor the development of new, unexpected emotions, such as *anger* and *indignation* (in Chapter III):

> 'Mine is a long and a sad tale!' said the Mouse, turning to Alice, and sighing.
>
> 'It is a long tail, certainly,' said Alice, looking down *with wonder* at the Mouse's tail; 'but why do you call it sad?' And she kept on *puzzling* about it while the Mouse was speaking.
>
> 'You are not attending!' said the Mouse to Alice severely. 'What are you thinking of?' [...] 'You insult me by talking such nonsense.'[18]

Alice's wonder while examining the Mouse's tail doesn't come here from amazement, but from incomprehension. Consequently, rather than in astonishment, the "surprise of non-sense" will resolve itself in anger. A similar inversion is that which allows going from desire to *regret*. "Alas! it was too late to *wish* that!"[19] The expectation is fulfilled,[20] but the fulfillment largely exceeds what was expected (e.g. Alice

---

[16] Carroll, *Alice*, 25.

[17] Carroll, *Alice*, 27.

[18] Carroll, *Alice*, 33–34.

[19] Carroll, *Alice*, 38.

[20] We could also mention here the anticipation of someone else's surprise, when Alice says, speaking of the White Rabbit (chapter IV): "*'He took me for his housemaid,' she said to herself as she ran. 'How surprised he'll be when he finds out who I am!'*" See Carroll, *Alice*, 37.

starts growing but cannot stop or control the process.). It even seems that desire has discovered here its guilt: "I almost *wish* I hadn't gone down that rabbit-hole – and yet – and yet – it's rather *curious*, you know, this sort of life! I do *wonder* what can have happened to me! When I used to read fairy-tales, I *fancied* that kind of thing never happened, and now here I am in the middle of one!"[21] In Alice's regret one can even sense a certain amount of *fear* of what might still happen and of the fiction-like events of Wonderland. The uncertainty of a return to normalcy and restoration familiarity generates the clear feeling of an impasse. And though, within this very state of despair, we assist to a new emergence of surprise (the pebbles thrown at Alice by the White Rabbit and his housemaid undergo a peculiar metamorphosis): "Alice noticed with some *surprise* that the pebbles were all turning into little cakes as they lay on the floor, and a bright idea came into her head. 'If I eat one of these cakes,' she thought, 'it's sure to make some change in my size; and as *it can't possibly* make me larger, *it must* make me smaller, *I suppose*.'"[22] Alice's reasoning here shows that she has adopted the thought criteria of Wonderland and given up those of an experience that does not contradict logic. The new surprise of the transformed pebbles gives her back the strength necessary to desire again and to hope the improbable from an object that could contribute to the fulfillment of her desire: She now knows that her height will change each time she eats or drinks something; she doesn't hold the key to the process but, as she finds herself excessively enlarged, she can hope that the little cakes will help her regain her normal height. This happens indeed, and the fulfilled expectation once again changes despair into satisfaction.

The encounter with the caterpillar (Chapter V) generates though a new interrogation wave concerning personal identity:

> "Who are you?" said the Caterpillar. This was not an encouraging opening for a conversation. Alice replied, rather *shyly*, "I – I hardly know, sir, just at present – at least I know who I was when I got up this morning, but I think I must have been changed several times since then." "What do you mean by that?" said the Caterpillar sternly. "Explain yourself!" "I can't explain myself, I'm afraid, sir" said Alice, "because *I'm not myself*, you see." "I don't see," said the Caterpillar. "I'm afraid I can't put it more clearly," Alice replied very politely, "for I can't understand it myself to begin with; and being so many different sizes in a day is very confusing." "It isn't," said the Caterpillar. "Well, perhaps you haven't found it so yet," said Alice; "but when you have to turn into a chrysalis – you will someday, you know – and then after that into a butterfly, I should think you'll *feel it a little queer*, won't you?" "Not a bit," said the Caterpillar. "Well, perhaps your feelings may be different," said Alice; "all I know is, it would *feel very queer to me*." "You!" said the Caterpillar contemptuously. "Who are you?" Which brought them back again to the beginning of the conversation.[23]

Alice can only respond shyly to the Caterpillar's abrupt question, and she will remain calm and polite while her interlocutor is completely non-empathic and insensitive to her identity doubts and to her feelings of strangeness. However, confronted with such incomprehension, Alice is progressively filled with dissatisfaction and develops a state of irritation that appears as the emotional correlate of

[21] Carroll, *Alice*, 39–40.

[22] Carroll, *Alice*, 43.

[23] Carroll, *Alice*, 47.

contradicted meaning intentions and of *contradicted desire*: "Alice said nothing: she had never been so much contradicted in her life before, and *she felt that she was losing her temper*".[24] Thus, instead of being astonished by her encounter with the talking Caterpillar, Alice becomes angry because of its non-responsive, meaningless speech.

In Chapter VI, the loss of the marks of normality and familiarity is intensified by the encounter with the Duchess and the smiling, appearing-disappearing Cheshire Cat. The exposure to nonsense becomes so radical that it confronts Alice with the hypothesis of (generalized) madness:

> 'In that direction,' the Cat said, waving its right paw round, 'lives a Hatter: and in that direction,' waving the other paw, 'lives a March Hare. Visit either you like: they're both mad.' 'But I don't want to go among mad people,' Alice remarked. 'Oh, you can't help that,' said the Cat: 'We're all mad here. I'm mad. You're mad.' 'How do you know I'm mad?' said Alice. 'You must be,' said the Cat, 'or you wouldn't have come here.'[25]

Madness as a general loss of sense and coherence is presented as unavoidable by the Cheshire Cat. In parallel, Alice starts to *get used to the marvelous*, thus neutralizing it as such: "'Do you play croquet with the Queen to-day?' 'I should like it very much,' said Alice, 'but I haven't been invited yet.' 'You'll see me there,' said the Cat, and vanished. Alice was not much *surprised* at this, she was getting so *used to queer things happening*."[26] Too much of the extraordinary kills the extraordinary; Wonderland cannot remain a land of the marvelous if marvels are everywhere. But the mechanisms of surprise are more complex than what can be dissolved in expectations. The Cat suddenly appears again and, when he disappears for the second time, Alice cannot help expecting to see it reappear. But the expectation is deceived, thus suggesting that once again Wonderland will never entirely fall under the law of desire and will only obey this law in order to subsequently disobey it. And yet Alice will try to impose her desire and to control the Cat's apparitions:

> [...] 'and I *wish* you wouldn't keep appearing and vanishing so *suddenly*: you make one quite giddy.' 'All right,' said the Cat; and this time it vanished quite slowly, beginning with the end of the tail, and ending with the grin, which remained sometime after the rest of it had gone. 'Well! I've often seen a cat without a grin,' thought Alice; 'but a grin without a cat! It's *the most curious thing I ever saw in my life*!'[27]

If the Cat (and through him Wonderland) obeys Alice's desire, it is only in order to launch a challenge to logic, thus indicating that fulfilled desire and verified meaning cannot coexist here: the Cat disappears, not suddenly but progressively, so that in the end Alice can only see his grin. The sheer contradiction of the fact that the attribute (the grin) can subsist without the substance (the Cat), or the part without the whole, is a genuine and most powerful source of surprise and astonishment for

---

[24] Carroll, *Alice*, 51.

[25] Carroll, *Alice*, 62–63.

[26] Carroll, *Alice*, 64.

[27] Carroll, *Alice*, 64–65.

Alice, insofar as it clearly goes against the basic rules of meaning. The criteria of the impossible are completely distorted here and have lost their force over experience.

The tea-party hosted by the Mad Hatter (Chapter VII) increases and aggravates the exposure to nonsense. Communication is here simply impossible: "Alice felt dreadfully *puzzled*. The Hatter's remark seemed to have *no sort of meaning* in it, and yet it was certainly English.".[28] And once again, the emotion caused by incomprehension and meaninglessness is *anger*: "'Have some wine,' the March Hare said in an encouraging tone. Alice looked all round the table, but there was nothing on it but tea. 'I don't see any wine,' she remarked. 'There isn't any,' said the March Hare. 'Then it wasn't very civil of you to offer it,' said Alice *angrily*.'"[29] Such negative emotions will become dominant in the next chapters, especially along Alice's interaction with the Queen. Alice arrives on the Queen's cricket field once again *accidentally* (and yet the Cheshire Cat had predicted to her that he will see her there). The encounter with the Queen's gardeners that paint white roses in red prove to Alice that her capacity for astonishment and surprise is still intact. More importantly, this is also valid for her capacity to surprise herself by daring to *denounce nonsense* even when threatened with death: "The Queen turned crimson with *fury*, and, after glaring at her for a moment like a wild beast, screamed 'Off with her head! Off–' '*Nonsense*!' said Alice, very *loudly and decidedly*, and the Queen was silent".[30] What is surprising here is no longer the confrontation with nonsense, but the ability to fully measure it and to inscribe it within clear limits.

However, becoming aware of nonsense is sufficient neither to stop it nor to restore familiarity and meaningfulness. Alice finds herself over and over in situations that launch new challenges to logic. For instance, the (mereological) problem of the "grin without a Cat" returns in the shape of the problem (disputed between the King and the executioner) of the bodiless head that has to be beheaded: "The executioner's argument was, that you couldn't cut off a head unless there was a body to cut it off from: that he had never had to do such a thing before, and he wasn't going to begin at his time of life. The King's argument was, that anything that had a head could be beheaded, and that you weren't *to talk nonsense*."[31] The King's logic is here a stiff logic, insensitive to context and to the conditions of meaning, and it consequently generates nonsense. It is somewhat the opposite of the moral that the Duchess randomly attributes to one of her stories (Chapter IX): "*Take care of the sense*, and the sounds will take care of themselves."[32] The Duchess herself does not respect this moral, given her complete carelessness for semantic coherence. It is thus under the form of an absurdly placed moral that the imperative of sense and meaningfulness is pronounced within the Carrollian Wonderland. And in spite of their absurdity, the Duchess's moral creates a legitimate expectation in Alice's

[28] Carroll, *Alice*, 69.

[29] Carroll, *Alice*, 66. Cf. ibid., 73: "'Take some more tea,' the March Hare said to Alice, very earnestly. 'I've had nothing yet,' Alice replied *in an offended tone*, 'so I can't take more.'"

[30] Carroll, *Alice*, 78–79.

[31] Carroll, *Alice*, 84.

[32] Carroll, *Alice*, 87.

mind; at the end of a new story, Alice believes she will hear a new moral, and her great surprise comes from the deception of this meaningless expectation, for the Duchess loses her voice in front of the unexpected apparition of the Queen: "But here, to Alice's *great surprise*, the Duchess's voice died away, even in the middle of her favourite word 'moral'."[33]

Furthermore, in Chapter X (the lobster quadrille), Alice herself becomes voiceless and silent because of far too many equivocal situations concerning meaning and meaning intentions: "Alice said nothing; she had sat down with her face in her hands, *wondering if anything would ever happen in a natural way again*," and her silence seems to contain a kind of emotional dullness, quite distant from the curb of astonishment and anger. The trial of the cake theft arises though her curiosity, and her last great surprise in Wonderland will be that of her convocation as a witness in this trial: "Alice watched the White Rabbit as he fumbled over the list, *feeling very curious* to see what the next witness would be like, '– for they haven't got much evidence yet,' she said to herself. *Imagine her surprise*, when the White Rabbit read out, at the top of his shrill little voice, the name 'Alice!'."[34]

In this trial simulation that raises Alice's curiosity in spite of its absurdity, semantic inconsistency reaches a climax when her deposition – related to an event that she has not witnessed – gains the status of a crucial testimony. Sense is not mandatory here and contradiction is entirely allowed: "Some of the jury wrote it down 'important,' and some 'unimportant.' Alice could see this, as she was near enough to look over their slates; 'but it doesn't matter a bit,' she thought to herself."[35] The accusation of nonsense not only remains completely unnoticed, but becomes fully redundant. Meaninglessness is not a problem in this context and is accepted as such when, for instance, Alice states her opinion regarding the poem invoked as testimony by saying: "'I don't believe there's an atom of meaning in it.' The jury all wrote down on their slates, 'She doesn't believe there's an atom of meaning in it,' but none of them attempted to explain the paper. 'If there's no meaning in it,' said the King, 'that saves a world of trouble, you know, as we needn't try to find any'."[36] The indifference to sense culminates here with the absurd order, dictated by the arbitrary and illogical will of the Queen, that the sentence be pronounced before the verdict:

> 'Let the jury consider their verdict,' the King said, for about the twentieth time that day. 'No, no!' said the Queen. 'Sentence first – verdict afterwards.' '*Stuff and nonsense*!' said Alice loudly. 'The idea of having the sentence first!' 'Hold your tongue!' said the Queen, turning purple. 'I won't!' said Alice. 'Off with her head!' the Queen shouted at the top of her voice. Nobody moved. 'Who cares for you?' said Alice, (she had grown to her full size by this time.) 'You're nothing but a pack of cards!'[37]

[33] Carroll, *Alice*, 89.

[34] Carroll, *Alice*, 111.

[35] Carroll, *Alice*, 114.

[36] Carroll, *Alice*, 117.

[37] Carroll, *Alice*, 118–119.

This ultimate denunciation, more violent than the previous, is also strong enough to cause Alice's awakening – an ending that seems to aim at restoring what narrative fiction has constantly subverted: the clear and rigid separation between reality and nonsense. Through Alice's awakening, the illogical domain of (literary) fiction is reabsorbed into the familiar world governed by logical legality. A victory of sense over nonsense, of meaning over meaninglessness? It is difficult to conclude in this direction, knowing that in Carroll's world (like in general), the time of being awake is only an *interruption*.

***

A few words of conclusion: If surprise rhythms, by its diverse figures and occurrences, Alice's adventures in Wonderland and their narrative construction, it doesn't strictly correspond to the experience of the marvelous and does not always bare the emotional color of astonishment and amazement. Negative emotions, such as disappointment and anger, are equally essential to surprise insofar as the experience of nonsense is as present in *Alice* as that of the marvelous. The double logic of sense and desire is thus the measure of surprise in the Carrollian Wonderland, explaining the variety of emotional responses to (dis)satisfied desire and to lack of meaning. In the light of nonsense, the intertwining of the cognitive and the emotional appears in its complexity; the surprise caused by the lack of meaning is not univocal, its valences are diverse, such that its connotations can go from astonishment to fear and anger. While the *marvelous* becomes *absurd*, the increasing amount of incomprehension will thus favor the development of new, unexpected emotions, like anger and indignation.

Part V
Surprise in Linguistics

Surprised? Why? The Expression of Surprise in French and in English: An Experimental Approach

Pascale Goutéraux

Abstract The psycholinguistic research reported in this paper posits a chaining of cognitive phases typical of any surprise episode and the inter-relatedness of emotional and rationalizing linguistic features. An experimental approach has been used to explore the reasons why the presentation of aesthetic visual 'sources'- a series of works of art- can trigger surprise reactions, hence the title: *Surprised? Why?* Although surprise is spontaneously expressed and described in fragmented, disorganized, apparently haphazard verbal forms, we have attempted to reconstruct the cognitive scripts underlying the discourse of surprise, by focusing on linguistic manifestations of disconnection, emotional response and adjustment, as three main phases of the cognitive schema of surprise. This work-in-progress is based on 25 recordings and transcripts of spontaneous reactions and emotive self-reports by French and English monolinguals, bilinguals and advanced learners of English. It is part of a larger corpus under completion at University Paris-Diderot, France, within the framework of *Surprise, a Vector for Enlarged Cognition*, an interdisciplinary project funded by the Research National Agency (French ANR). The study aims at uncovering invariant and variable psycholinguistic features in the surprise discourse of native and non-native speakers of English and French.

Keywords Expression of surprise · Psycho-linguistics · French · English · Cognitive strategies · Emotions · Experimental task · Works of art

Most psychologists of emotion (Plutchik 1980; Damasio 1994; Ekman 1992) have put surprise in the category of emotion. Discourse specialist, Cosnier (2006), refers to surprise as either a primary emotion like fear, disgust, anger, pleasure, joy, sadness or a "mixed" one, resulting from a blend of basic emotions, all of which are short-lived and caused by unexpected events. Linguistic research has frequently treated specific lexical items of surprise as emotion predicates (Johnson-Laird and Oatley 1989; Goossens 2005; Novakova et al. 2012) among others. However, the view of surprise as a full-fledged emotion has been challenged by

P. Goutéraux (✉)
University of Paris Diderot, Sorbonne Paris Cité, CLILLAC-ARP, EA 3967, Paris, France

N. Depraz, A. J. Steinbock (eds.), *Surprise: An Emotion?*, Contributions To Phenomenology 97, https://doi.org/10.1007/978-3-319-98657-9_10

cognition-oriented research. The psycholinguistic research reported in this paper posits a chaining of cognitive phases typical of any surprise episode and the interrelatedness of emotional and rationalizing linguistic features. An experimental approach has been used to explore the reasons why the presentation of aesthetic visual 'sources'[1] – a series of works of art – can trigger surprise reactions, hence the title: *Surprised? Why?* Although surprise is spontaneously expressed and described in fragmented, disorganized, apparently haphazard verbal forms, we have attempted to reconstruct the cognitive scripts underlying the discourse of surprise, by focusing on linguistic manifestations of disconnection, emotional response, and adjustment, as three main phases of the cognitive schema of surprise. This work-in-progress is based on 25 recordings and transcripts of spontaneous reactions and emotive self-reports by French and English monolinguals, bilinguals, and advanced learners of English. It is part of a larger corpus under completion at University Paris-Diderot, France within the framework of *Surprise, a Vector for Enlarged Cognition*, an interdisciplinary project funded by the Research National Agency (French ANR). The study aims at uncovering invariant and variable psycholinguistic features in the surprise discourse of native and non-native speakers of English and French.

1 Surprise: A Unique Combination of Cognitive and Emotional Features

As early as 1987, Ortony et al. considered surprise as a cognitive state caused by some unexpected event that can merge into positive and negative surprise, depending upon the quality of the stimulus (p. 32). The fields of psychiatry and psycho-physiology have also stressed the physiological 'shock and startle' characteristics of surprise (Desmidt, this volume): accelerated heart rate or sweating (skin conductance increase); the 'startle' manifestation also refers to bodily movements due to an instinctual and eventually surprised and fearful reaction to a stimulus, images and memory images with aversive content (Lang et al 1990), for instance in experiments using the International Affective Picture System (IAPS) database (Bublatzky et al. 2013). A more complex picture of the cognitive and affective phenomenon of surprise has recently emerged with the integration of the concept of emotional schema or scenario. Appraisal psychologists added a cognitive dimension to the affect-oriented description of emotions by contending that any emotion is part of a cognitive script or schema composed of a triggering event, a cognitive and emotional assessment of this event, and physiological, behavioral and verbal reactions, including subjective assessments corresponding to what one thinks of or says about the event (Syssau 2006). Some cognitive scientists have described surprise as a general affective answer to an unexpected event rather than a specific response (Stein and Hernandez 2007: 302):

> Some general affective responses, such as surprise, are elicited when inferences have been made about the unexpectedness or novelty of a precipitating event. Inferences about the

[1] The word 'source' (Plantin 2011) rather than 'stimulus' will be used to prevent a behaviorist interpretation of the phenomenon of surprise.

> novelty of events are critical to the experience of emotion and are produced by pattern recognition and pattern matching procedures that are continuously carried out in states of wakefulness and sleep.

The uniqueness of surprise, resulting from a cognitive discrepancy between a familiar or well-internalized set of mental representations and an event perceived as new and disturbing, was highlighted by the *Cognitive Evolutionary Model* (Meyer et al. 1997: 253):

> If a discrepancy between schema and input is detected, surprise is elicited, schematic processing is interrupted, and a more effortful, conscious, and deliberate analysis of the unexpected event is initiated.

Reisenzein (2000: 262) refined the model by inserting an experiential phase – a feeling of surprise and non-reflective awareness – before the behavioral manifestations (bodily and verbal ones) and the more analytical (re)-adjustment phase:

> Like all emotions, surprise has a conscious or subjective aspect. This aspect manifests itself in two properties that seem to be central for conscious states generally: phenomenal character and immediate awareness: On the one hand, *it is like something* to be surprised or, as emotion psychologists usually say, it *feels* in a particular way to be surprised. On the other hand, *one is immediately aware* of one's surprise experience; that is, one is—in a non-focal, pre-reflective way—aware of one's experience without, apparently, having to draw any inferences.

This research has adopted a broad view of surprise as a dynamic process and endeavored to reconstruct the cognitive phases of a surprise episode by analyzing the spoken discourse of subjects expressing spontaneous surprise and reporting upon their subjective experience. Firstly, verbal manifestations of disruption and adjustment, these invariant elements constitutive of the cognitive scheme of surprise, were retrieved and accounted for, and secondly, we analyzed linguistic combinations displaying associated emotional facets of the surprise experience in terms of intensity and valence, two dimensions that we believe enhance the subjective experience of surprise.

1.1 Working Hypotheses

The first hypothesis that was examined is that any surprise experience is characterized by a core invariant, the subjective feeling of a disruption of a cognitive script deeply-rooted in individual expectations based on cultural and affective mental representations and personal episodic memories. The sense of discrepancy and destabilization may be implicit first and then explicitly verbalized as the experiencer becomes (suddenly or gradually) aware of the breach in her/his own internalized representations caused by a sudden unexpected event. Cognitive disconnection may take the shape of a single brief disruption or a series of surprise micro-episodes as the subject revisits the triggering source mentally and/or visually.[2] As stated above,

[2] Cf. the technique of the elicitation interview was developed by Pierre Vermersch (1994), and is currently used by phenomenologist Natalie Depraz (this volume).

we will also hypothesize that as an invariant pivotal point of the surprise process, cognitive disruption is often associated with various emotions such as fear, disgust, sadness, joy or pleasure – expressed by an array of linguistic forms. Whether these emotions are only possible outcomes or are fully integrated in the process (surprise segueing into emotions) is still a matter for discussion.[3] The current psychological research on surprise seems to have evolved towards a concept of surprise devoid of any strongly positive and negative valence (Ortony et al. 1988; Ekman 1992). However, lexical studies have also uncovered surprise lexical networks semantically marked by two typical emotional dimensions: intensity and valence.[4] For instance, Novakova et al. (2012: 1020–1021) underlined the frequent use of adverbial intensifiers modifying surprise adjectives and verbs, in French, English and Spanish written texts. They also identified a positive or negative polarity in surprise-laden lexical phrases referring to the source or to the experiencer's resulting state of mind. Thus, the data collected in this experimental research has been examined to see if it can support various scenarios of surprise, declined as an intense emotional experience, a neutral disruption of previously-held beliefs or a blend of cognitive disruption and emotional perception.

Apart from the lexemes derived from the lemma *surprise*, *mirativity* – surprise in linguistic terms – is not expressed by any specific linguistic marker (a particle, etc.) in French or English, the two languages used in this experimental study,[5] and yet a sense of surprise may emerge from the speakers' utterances. We have hypothesized that surprise – either spontaneously expressed or described – is verbalized through a network of both explicit and indirect markers. It can be clearly identified and named: "*I am surprised*", "*it's surprising*", "*c'est surprenant*", as the speaker describes his/her own state of mind or attributes a surprising quality to the source. Most often surprise discourse results from the mapping of various categories of linguistic markers:

- lexical items, synonyms and derivatives,
- expressive markers such as false starts, silent and filled pauses, bouts of laughter, hesitations bordering on speechless stupefaction and interjections,
- syntactic patterns like exclamations and questions as well as epistemic modal forms, surprise-laden lexical items or complex metaphorical and metonymic phrases.

This view of the discourse of surprise as a network of markers tallies with the pragmatic approach to affect that differentiates 'emotional discourse', as a spontaneous reaction, from 'emotive discourse', by which the subjects (or a third person)

[3] *Surprise an Emotion*? A colloquium organized by A. Steinbeck and N. Depraz, September 25-27, 2013, Southern Illinois University, Carbondale, Illinois, USA.

[4] 'Valence' (in its psychological sense) has been chosen over 'polarity' as it refers to a scale of positive, negative and neutral traits.

[5] "Mirativity refers to the linguistic marking of an utterance as conveying information that is new or unexpected to the speaker" (Delancey 2001: 369–370). In this seminal article, Delancey reviewed a number of languages featuring specific markers of surprise.

describe an affective experience and comment upon it (Caffi and Janney 1994; Plantin 2011). However, applying 'emotional' and 'emotive' labels to the spoken discourse of surprise presents some caveats. Firstly, the disorganized nature of the spoken flow tends to blur the distinction between spontaneous expression versus description and comment; they seldom appear sequentially, and surprise utterances often display a mix of expression, description and rationalization stances; for instance, "unbelievable!" is expressive (exclamation mark), descriptive (surprise-laden adjective applied to the source), and rationalizing since the feeling of surprise is attributed by the speaker to a breach in his common beliefs. Secondly, utterances that a third person could identify as occurrences of surprise by connecting various speech markers may never be explicitly labeled as such by the speaker; conversely, naming a source as 'surprising' is not always preceded by any spontaneous verbal manifestation (one can simply internalize surprise or be rendered speechless by shock). Finally, we need to differentiate surprise words (a limited set derived from the word surprise) from surprise-laden words (for instance, *different, strange, weird* or French 'impressionnant'); this concept is modeled upon the distinction between emotional words and 'affect words indirectly expressing emotion' (Ortony et al. 1987: 341) or emotion and emotion-laden words (Altarriba 2006; Pavlenko 2008).

1.2 Corpus Study: Presentation of the Experiment

These hypotheses have been tested by analyzing the responses of twenty-five informants to a series of works of art as part of a psycholinguistic task designed to collect a corpus of reactions and self-reports of surprise by French and English students from the English Studies Department at Paris-Diderot University, France.[6] Each student was shown – on a computer screen – a series of twelve anonymized paintings or sculptures chosen to trigger affect and surprise[7] reactions. These aesthetic objects – ranging from the Middle Ages to the contemporary art scene – were selected to offer multiple opportunities to trigger surprise, as well as a whole range of aesthetic and emotional reactions. The pictures were presented in a pre-determined order. Some were expected to induce surprise reactions while others operated as distractors (a bouquet of flowers, a nice country landscape or a woman holding a letter). The order was sometimes manipulated to maximize the contrast between two aesthetic sources and enhance the potential reaction of surprise by thwarting the viewers' expectations as to what they would see next: for instance, Gustave Dore's

[6] The specific discourse of foreign language learners and bilinguals regarding the expression of surprise is part of another work in progress to inform research on Second Language acquisition. Linguistic proficiency as a dependent variable (Dewaele 2010) is expected to play a major part.

[7] Two different series were alternated to prevent the subjects, who had signed a confidentiality agreement, from communicating on the contents. They were told that the experiment was about language and cognition but not that it revolved around the expression of surprise.

Vision of Paradise (1861) was followed by Joachim Beuckelaer's *Slaughtered Pig* (1563); Jan Mabuse's *Danae* (1527) preceded Paul Rebeyrolle's *Implosion* (1994).

The first part of the experiment consisted in recording the viewer's spontaneous reactions during a short interval (no longer than 60 s) between each viewing; in the second part, a native speaker conducted a semi-directed interview of the students in English or in French: the interviewees were asked to remember the surprise episodes they had experienced and talk about the most marked ones; then they were shown the pictures again to confirm (or disconfirm) their initial reaction of surprise. They named the emotional valence – if any – they associated with the pictures and discussed the reasons why they had felt surprised (cf. Appendix). The corpus thus far includes the transcripts of twenty-five speakers: four native speakers of English, three bilinguals who immigrated to an English-speaking country at a very early age, eight advanced learners of English, and ten native speakers of French.[8] Such a small-sized, linguistically heterogeneous sample warrants a qualitative analysis, and both the results and our interpretations will have to be put in perspective by analyzing a higher number of productions. We did not adopt a behaviorist stance by having recourse to a simplistic causative scheme of surprise (X event causes surprise), although we must acknowledge some effects were predictable. Focusing on the interactions between the specificities of the source that disrupted the mental and linguistic expectations of the subjects and the actual discourse the speakers produced enabled us to elicit more fine-grained characteristic features of a surprise scenario.

2 Surprise Discourse: At the Crossroads Between Cognitive Schemas and Linguistic Networks

Reisenzein's sophisticated script was used as a guideline for the exploration of the linguistic marking at different stages of the surprise process.[9]

[8] 72 subjects have participated in the experiment so far (2018), *i.e.* 12 native speakers of English, 10 native speakers of French, 12 French–English bilinguals and 38 advanced learners of English. The aim is to elicit core universals common to all speakers as well as cognitive and linguistic variations depending on the language spoken and the native, non-native or bilingual status of the speakers.

[9] "Surprise-eliciting events initiate a series of mental processes that begin with the appraisal of a cognized event as exceeding some threshold value of unexpectedness or schema-discrepancy, continue with the interruption of ongoing information and, simultaneously, the interruption of ongoing information processing and the reallocation of processing resources to investigation of the unexpected event, and culminate in the analysis and evaluation of this event plus […] immediate reactions to the event and/or the updating, or revision of the schemas that gave rise to the dis-comfirmed expectation"(Reisenzein 2000).

2.1 *The Linguistic Marking of Cognitive Disconnection*

When presented with an aesthetic object they found surprising, the speakers frequently expressed a sense of unexpectedness and verbalized their awareness of the discrepancy between their expectations and the novelty of the event; they investigated the aesthetic object, occasionally verbalized their inability to make sense of it and/or attempted to revise and update their representations and beliefs. Surprise utterances appear randomly, either in the temporal slots allocated to spontaneous reactions in the first part or during the subsequent interview. Actually, the dispersion of surprise utterances made it necessary to screen the totality of the transcripts when searching for surprise or surprise-related markers as in Example 1

(1) Emphi022: French Speaker (interview)[10]

Celle-là était était assez surprenante ouais mais c'est ce qui fait aussi du coup que ça m'a ...que j'ai trouvé ça très amusant [...] parce que y avait le décalage avec quelque chose de pratique et quelque chose qui de toute évidence n'est pas pratique en réalité quoi !

(*This one was was rather surprising yeah but that's also why it made me why I found it very funny [...] because there was this discrepancy between something practical and something which obviously is actually not practical at all.*)

The discrepancy between the folk vision of a typical cup and saucer set (*quite practical*) and its furry representation by Oppenheim[11] is identified as a cause of surprise (*something which is not practical at all in fact*). The disruption of beliefs is expressed by several linguistic means: the French i*mparfait* + an explicit adjective of surprise attributed to the source (surprenant) and a causative clause including a marker of disruption (décalage). In Example 2 (same source), surprise is not named, but several markers concur to support the idea that the speaker is indeed surprised.

2) Emphi006: Advanced Learner of English (reaction part)

Well we are used to a certain image as regards the utensils and objects we use every day but here (erm) it is represented in a manner I would never have expected and (em) I find it quite funny actually. (erm) because I would never have thought of that [...]

In Examples 1 and 2, surprise results from the disruption of commonly shared representations of everyday life objects (*we are used to a certain image*). The third conditional in Example 2 refers to an implied past condition that did not happen and a resulting event that should not have occurred: *I would never have expected/I would never have thought of that*. And yet, the artist did make a furry cup and saucer, hence the speaker's feeling of amused surprise (*quite funny*). Similarly, Example 3 shows a conscious self-reconstruction of the reasons why two other paintings had a surprising effect upon the speaker:

---

[10] The two parts of the experiment are respectively identified as 'reaction' part and 'interview' part. In the quotations, periods are used to encode silent pauses (one period roughly corresponds to one second, two to two seconds, etc.). Translations in English have been provided in italics whenever the original quotation was in French, and the examples in English are in italics as well.

[11] Meret Oppenheim, *Fur Breakfast*, 1936.

(3) Emphi002: Native Speaker of English (reaction part)
like I would never have (er) have thought to to paint a a woman in a womb[12] *or I mean perhaps that's just me or the carcass I would never have done*[13] *that I mean I wouldn't have wanted to to look at a thing like that more than necessary so that's quite surprising me.*

Besides the use of third conditionals (3 tokens), the lexical verb *surprise* modified by the intensifier *quite* takes on an aspectual and modal dimension (*be* + *surprise* + *ing*). Aesthetic choices deemed impossible in the speaker's experience are contradicted by the very paintings she has just been shown, hence her endorsement of her surprise experience as an ongoing process: *that's quite surprising me.* She confirmed her stance by rating *Danae* (she loved it) and *Slaughtered Pig* (she hated it) as the two most surprising pictures of the series in the subsequent interview.

2.2 The Disruption of Aesthetic Representations

Because of the aesthetic framework of the experiment, surprise speech often ensues from the disruption of the speakers' expectations as to what art is supposed to be. Surprise may occur spontaneously in reaction to one specific painting and/or be explicitly stated in the interview. 'Interview speech' actually exhibits few spontaneous surprise markers. Grammatical tenses (French *imparfait* or English preterit) often mark the disconnection between one's aesthetic representations and the visual object presented as art.[14] The speakers also tend to rationalize their reaction: in Examples 4, 5 and 6, the subjects redirect the cause of surprise from the aesthetic object *per se* to the artist's intentions as they perceive them, perhaps to distance themselves from the emotional impact of the content of the picture.

(4) Emphi017: Native Speaker of French (reaction part)
Mais oui ça m'a surpris je ne m'attendais pas à ce que quelqu'un ait peint euh[15] [...]
(*Yes, it surprised me I was not expecting anyone to have painted er [...]*)

Example 5 displays markers of disruption through the broken speech flow (*yes, it's really er, er*) followed by negated expectations (*was not expecting*), generalization by shifting pronouns, from singular *I* to plural *we* (*I wasn't expecting to see that/we have the impression*), a slightly negatively valenced word (surprise-laden *weird*), and finally the rationalization of the weirdness she experienced when looking at the painting (*like the work of a child*).[16]

(5) Emphi001: Advanced Learner of English (reaction part)

---

[12] Gustav Klimt, *Danae et la pluie d'or*, 1907–1908.

[13] Joachim Beuckelaer, *Slaughtered Pig*, 1563.

[14] The initial directive to the students starts this way: 'you're going to see a series of works of art ...'.

[15] Lucian Freud's *Naked Man*, 1992.

[16] Ken Kiff, *Green Man*, 1977.

*Yes it's really (er) (er) I I wasn't *expected expecting to see that (er). It's really weird because we have the impression that it's like (er) the work of a child almost*

When one's aesthetic beliefs are repeatedly challenged by several pictures, the discourse of surprise becomes less emotional and eventually takes on a more intellectualized character, as in Example 6.

(6) Emphi025: Native Speaker of English (interview)

(er) I think I first because I'm working at university I was expecting to see some works of art and that (er) I'd want to interpret in some way and (er) I found the pictures quite (er) strange often quite disturbing very varied (er) but that was quite interesting I like (er) [...] (erm) surprising I guess I was surprised by (er) they they seem to come from very different periods or different backgrounds.

The speaker claims her academic environment favored conventional artistic expectations and was conducive to assessing and interpreting the pictures rather than reacting to them. Valence is low-key, ranging from vaguely negative items (*strange, disturbing*) to neutral (*surprised, surprising, and different*) or moderately positive (*interesting*) ones. Surprise is explicit (2 lexemes) and reinforced by intensifiers *quite* (3 tokens) and *very* (2 tokens) modifying surprise-laden adjectives (*varied*).

2.3 Disconnection, Puzzlement and Re-adjustment Action Phase

As previously mentioned, the disruption of cognitive and affective representations often induces a state of confusion followed by "a more effortful, conscious, and deliberate analysis of the unexpected event" (Reisenzein 2000) and a phase of reorganization or adjustment. The speakers correlate their feeling of surprise and their inability to make sense of what they've just seen. They report being *puzzled, confused,* or in French 'intrigués, déconcertés, perplexes', or they refer to the source as *puzzling, confusing*. In Example 7-1, the speaker responds spontaneously to Gustave Doré's *Vision of Paradise*[17] by implicit markers of surprise (laughter, filled pause *er*), followed by a series of surprise and surprise-laden lexical items and introspective comments regarding her feeling of disorientation.

(7-1) Emphi023: Native Speaker of English (reaction part)

(laughs) (er) so this is one of those (er) pictures or or paintings that that unsettles me because I don't know what I'm supposed to see and it sort of takes me out of my my position as a spectator (er) so it's it's surprising (er) my first reaction is a kind of a short laughter (er) and I think it's also bec I laugh because I'm unsettled because I don't know how I'm supposed to react to it.

The speaker, who later rated this picture as the most surprising of all, justified her short initial burst of laughter by the unsettling effect the picture had on her (*unsettles/*

[17] Gustave Doré, *Vision of Paradise*, 1861.

unsettled) and the sudden change of roles – from spectator to actor – she experienced; she also attributed it to the state of cognitive confusion she found herself in (*I don't know what I am supposed to see, I don't know how I am supposed to react to it*). At this stage, her speech did not exhibit any distinct emotional lexical feature nor did she try to rationalize her confusion. The negative valence of her reaction became explicit further on in the interview as she expanded on her physical rejection of the picture:

(7-2) (interview) *yeah I mean I my first reaction was like to turn away from it I don't like it but then perhaps the more because that's the one that made me the most (er) it was the most uncanny one for me (er) I this one it stood out because it was not it's not something I was familiar with it's the kind of work of art I perhaps don't understand and I'd rather just not even (er) look at it so that's the kind of thing that I sort of reject.*

Surprise is re-asserted by surprise-laden words (*uncanny, stood out*); initial puzzlement is updated and rationalized (*not something I am familiar with, the kind of art I perhaps don't understand*) and accompanied by metonymic phrases expressing physical rejection (*turn away from it, not even look at it*) and verbal negative stance (*I don't like it, I sort of reject*). Thus, extracts 7-1 and 7-2 show that the phases of the cognitive processing of a surprising event can be distributed over different temporal slots (reaction and interview) and also how dispersion is mirrored in speech by the production of fragmented and incomplete clauses (*but then perhaps the more because that's the one that made me the most (er) it was the most uncanny one*).

Similarly, Example 8[18] testifies to the revisiting of an episode of surprise, although the clause-order reflects neither the actual succession of temporal phases nor the sequencing of the cognitive processing. The speaker switches between tenses while reporting on his experience of surprise (past tenses), commenting the surprise episode (use of the present tense) and stating his current emotional state (*torn between two feelings*) during his second viewing of the painting.

(8) Emphi005: Late Bilingual[19] (interview):

Yeah I was surprised by the meat (erm) [...] this one was a little bit. yeah it caught me off guard cuz I wasn't expecting that kind of picture [...] well. I asked myself whether the fact that I thought it was very well painted made me like the picture because I kind of I'm as I said I'm quite repelled by that aspect but at the same time I appreciate good art good composition and skills so it was hard for me to have an emotional reaction that was specific because there's both the repelling side of it but also the attractive side of it [...] torn between the two feelings that I don't really like it but I sort of like it as well so yeah the ambiguity of it.

After recalling his surprise at seeing Beuckelaer's *Slaughtered Pig* (metonymically referred to as *the meat*), the speaker explains how he evolved from sudden surprise (*caught me off guard*) to conscious awareness and identification (*I wasn't expecting that kind of picture*). He then investigates the reasons why he was both negatively affected (*repelled*) and aesthetically attracted (*I appreciate good art,*

[18] Joachim Beuckelaer, *Slaughtered Pig*, 1563.

[19] This student spent several years in various English-speaking countries and qualifies as a late bilingual.

good composition and skills) and finally comes to terms with this contradiction (*torn between the two feelings*) and his ambiguous reactions (*don't really like it but sort of like it as well*).

We have found numerous other examples displaying a mix of puzzlement and emotion (phases two and three of the surprise scenario) but the speech flow seldom reflects the canonical order of the cognitive script. In Example 9-1, the interviewee has asked to see Bourgeois's *Janus Fleuri* again, a painting he remembers as surprising[20]:

(9-1) Emphi022: French Speaker (interview)

J'avais l'impression que…je je je mon cerveau sait cherche à savoir ce que c'est mais d'un autre côté (euh) ... Je préfère peut-être pas (laughs) savoir (laughs). ce que je regarde vraiment c'est. c'est étrangement en fait c'est étrangement gore.

(*I felt that... I I I my brain knows is trying to find out what it is but on the other hand (er) ... perhaps I'd rather (laughs) not know (laughs) what I am actually looking at it's. it's weirdly it's actually weirdly gore.*)

The speaker begins with recalling his initial experience (*j'avais l'impression de*) before delivering a 'live' report of his investigation processing (*mon cerveau cherche à savoir ce que c'est*) and concludes with a surprise-laden emotional phrase (*étrangement gore*), an emotional assessment that should have emerged earlier in speech. As can be seen in Example 9-2, the speaker's initial reaction was a disorganized string of bursts of laughter, filled pauses and interrupted stretches of discourse (*c'est vraiment/on dirait encore*):

(9-2) (euh) (laughs) c'est c'est encore une fois très particulier (laughs) c'est vraiment on dirait encore (euh) ... des parties d'être humain […]

(*(er) (laughs) it's again it's very special (laughs) it's really again (er) it looks like human body parts*)

Finally, Example 10 also shows cognitive dissonance reflected by a broken and disorganized series of markers and semantic associations. Although it does not present any surprise lexemes, the utterance can be interpreted as expressing surprise, mental confusion and rejection: two filled pauses (*er*), a short question (*What is it*?), a series of rapid negations (*no, no, no*) implying rejection, shifts in grammatical subjects (*no I no it looks I'm sorry*) and the final metaphorical association with a sexually transmitted disease:

(10) Emphi011: Advanced Learner of English (reaction)

(er) What is it? Okay (er) no no I no it looks I'm sorry, I just have to say it it looks like STD.

To sum up, Examples 7 to 10 illustrate various psycholinguistic configurations and confirm the partial overlapping – but not the superposition – of invariant cognitive surprise scenarios based on disconnection, reaction and action, and of a discourse characterized by a versatile network of explicit and implicit linguistic markers.

[20] Louise Bourgeois, *Janus Fleuri*, 1968.

3 The Lexical Expression of Surprise

The last part of this paper is devoted to the lexical component of the discourse of surprise and discusses the use of surprise-laden and associated emotion-laden words in the data and of metaphorical and metonymic phrases in French and English as well. Although the number of tokens produced for some types is occasionally mentioned, this is by no means a quantitative approach to the lexical expression of surprise in the corpus, given the limited number of informants. Also we have not differentiated between monolinguals, bilinguals, and learners in terms of proficiency but tried to show that the trends we are reporting here tend towards the universal.

3.1 Surprise and Surprise-laden Lexicon in Use

Of course, the lemma *surprise* takes on various linguistic forms: in English, adjectives *surprising* (*n* 10),[21] past participle *surprised* (*n* 7) and verb *surprise* (*n* 4); in French, adjectives *surprenant* (*n* 14), *surpris(e)* (*n* 1), verb *surprendre* (*n* 8). In the second part of the experiment, the interviewees had to answer a very direct question: *Were you surprised by some of the pictures*? But there was no need to repeat *I was surprised,* and they mostly answered *yes*, or nodded, or mentioned one or several pictures before expanding on their reactions. Lots of words are associated with the experience of surprise, either as synonyms or words designating one semantic property of surprise or an effect of the surprise process. Surprise is not always named but rather alluded to by a combination of surprise-laden and emotional words or by lexical imagery as shown by Example 11[22]:

(11) Emphi008: British Native Speaker of English (reaction part):
That's quite shocking (erm) kind of grotesque I also feel quite taken aback by it disturbing (erm) [...] mangled and twisted and yeah shocking.

Adjectives or past participles intensified by the adverb *quite* convey the physical impact of surprise (*taken aback*) or shock (*shocking*, *n* 2) and disruption (*disturbed, n* 1); they are intermingled with negative emotional adjectives attributed to the source: *grotesque, mangled* and *twisted.* Such effects of surprise as shock, startlement and repulsion are mentioned in the reaction part of the experiment although surprise is explicitly confirmed in the interview. Adjectives expressing the disturbing impact of the source's semantic contents on the experiencer's representations are numerous in the corpus, which supports the claim that in English and French affect discourse, adjectives are the major grammatical category in use (Novakova et al. 2012; Pavlenko 2008). The number of nouns is limited (*n* 6 in

[21] *n x* refers to the number of tokens of the same word form, in English or in French.

[22] Paul Ribeyrolle, *Implosion*, 1994.

English, *n* 9 in French – and verbs -*n* 4 in English and *n* 15 in French). English adjectives are *disturbing* (*n* 18) + *disturbed* (*n* 2), *unsettling/unsettled, taken aback*, etc. We also find a variety of forms expressing differentiation from commonly shared beliefs: *weird* (*n* 37), *strange* (*n* 24), *different* (*n* 7), *uncommon, unexpected, original, odd, bizarre, uncanny*. The same number of surprise-related types (53) was found in French utterances, with a gradient of disconnection markers attributed to the experiencer or to the source. They ranged from intensely negative to mildly positive or neutral ('perturbant, dérangeant, troublant, déroutant, intrigant, étonnant, déconcertée, intrigué'); as a sub-category, differentiation is the dominant property of the source ('différent, original, spécial, particulier, pas anodin, pas/peu commun, curieux, absurde, bizarre, étrange' – *n* 13 for the latter).

3.2 The Role of Intensity and Valence in the Surprise Lexicon

The discourse of surprise is often strengthened by the intensity of an associated emotion whether it is positive or negative, expressed or reported.[23] The pictures most frequently identified as eliciting strong surprise typically bring out reactions characterized by either a very positive valence (amusement, amazement) or a very negative valence (fear, disgust, shock or sadness). However, negative valence prevails (cf. Appendix) since out of twenty-four visuals (twelve per series), seven are repeatedly mentioned as quite surprising, and most are also assessed as negative. For instance, in the first series (Table 1), *Green Man* is quoted 10 times and elicits emotional reactions ranging from amusement to disgust. The mixing of horror and repulsion with surprise and shock is predictable in some cases. Actually, surprise has been identified as a recurrent component of fright in the literature (Valetopoulos 2013). *Slaughtered Pig* was described by 9 out of 12 speakers as 'horrible', 'scary' or a 'disgusting bloody carcass'. Bourgeois' *Janus Fleuri* (Table 2, 10 speakers) and Soutine's *Carcass of Beef*[24] (Table 2, 8 speakers) also consistently ranked as highly surprising and horrible, scary or disgusting. *Implosion* caused violent reactions of surprise in 12 out of 13 speakers (Table 2), a reaction emphasized by the contrast with the previous painting (Mabuse's *Danae*).

(12) Emphi007, Late Bilingual (interview):

It's just because they're really violent really and (er) ... yeah. and probably it's all the contrast between the two like the first one which is quite you know kind of a classicalish thing (er) which you would expect in you know to be hanging in a museum or something (er) and then just right right after that the the second one which is quite yeah …

In terms of intense positive valence, some paintings elicited amazement and wonder and surprise took on the more classical sense of admiration as a "sudden

[23] The variable interactions at play between the semantic properties of the aesthetic source and the individual history of a speaker may explain the variety of the subjects' appraisal reactions (work in progress).

[24] Chaïm Soutine, *Carcass of Beef*, 1925.

surprise of the soul" (Descartes, 1649).[25] For instance, *Vision of Paradise* (Table 1) inspired reactions of admiration and amazement in 8 out of 12 speakers.[26] Other aesthetic sources triggered more individualized positive reactions. Ford Maddox Brown's painting, *the Corn Field* (1850) (Table 2), was said to be *surprising* and *trippy*:

(13) Emphi010: Bilingual (reaction)
Oh I like this it's it's surprising it looks like Haitian stuff like Haitian (erm) what do they call it Haitian naïve yeah like those primitivist thingies I don't know it I kinda like it it's trippy.

Objects perceived as incongruous also induced amused surprise and perplexity, such as Jeff Lyons, *Illu*[27] (8 speakers) or *Fur Breakfast* (6 speakers) (Table 2). In most prototypical scenarios, a gradient of emotions contributes to increasing the intensity of the surprise experience.[28] Nevertheless, it seems difficult to disentangle surprise from other emotions and single out "pure" psychological and linguistic surprise manifestations. Besides surprise and surprise-laden lexemes, the lexical imagery of surprise is frequently built on the use of formulaic language, and the last paragraph focuses on the metaphorical and metonymic treatment of surprise in terms of bodily impact and kinesthetic apprehension.

3.3 The Physicality of Surprise: Kinesthetic and Metaphorical Discourse

In *Metaphor and Emotion, Language, Culture and Body in Human Feeling*, Slován Kövesces (2000: 5) describes the 'emotion of surprise' as a physical force entailing destabilization and loss of control. Like any emotional discourse, the discourse of surprise is replete with metonymies, whether a specific feature stands for the source or a mental or physical effect for the whole surprise process. Speakers obliquely refer to surprise as a blow delivered to them or as a shock they experienced. *Vision of Paradise* is conducive to the production of lexical imagery relative to a physical experience both in English and in French.

(14) Emphi014: French Speaker (reaction part)
Ce grand jaune comme un soleil qui vous frappe c'est choquant, enfin ça me choque.
(*This huge yellow thing like a sun striking you, it's shocking, I'm shocked*)

---

[25] Descartes, *Traité des Passions*, article 70: l'admiration est "une subite surprise de l'âme, qui fait qu'elle se porte à considérer avec attention les objets qui lui semblent rares ou extraordinaire."

[26] One subject reported being negatively disturbed by the dominant color (she didn't like yellow) and the religious connotations (as a result of her own family history).

[27] Jeff Lyons, *Illu*, 1996.

[28] This may also be partly due to the aesthetic nature of the sources, which tends to cause more affective reactions.

Surprise as shock and physical violence is a recurrent metonymy as in Examples 15 to 18, in which speakers are describing their experience of *Implosion*, either in the reaction part or the interview.

(15) Emphi010: Bilingual, (interview)
I was surprised by number two just because of like it's the sheer goriness of it […] it kind of takes you aback and (erm) I was like meta surprised.
(16) Emphi022: French Speaker (reaction part)
Çá vous prend aux tripes.
(*It's gut wrenching*).
(17) Emphi006: Advanced Learner of English (reaction part)
It's bang in the middle of the picture so it's just staring at you kind of thing.
(18) Emphi025: English Speaker (interview)
That one that was striking I think for the strange the mix the disgusting and also the beautiful and the pain in it […] yeah I found that really striking (*er*).

Regardless of their linguistic status as native speakers of French or English, bilinguals or advanced learners of English, all experience and report similar physical effects and verbalize them with a variety of lexical metonymies (*striking*, *takes you aback,* 'prend aux tripes', *bang*) reinforced by intensively valenced words (*sheer goriness*) or a mix of negative and positive nominalizations and nouns (*the disgusting, the beautiful, the pain*) expressing the highest degree of the emotion mingled with surprise. The subjects frequently exhibit physical effects of fear, as a result of the surprise experience.

(19) Emphi013: French speaker (reacting to *Slaughtered Pig*)
Ça met… ça donne froid dans le dos
(*It makes me…it's chilling*)
(20) Emphi018: French speaker (reacting to *Spider*[29])
Ça me fait des frissons un peu là, je ne suis pas à l'aise.
It sends shivers down my spine now I feel uneasy.

For another speaker the verbalization of spontaneous surprise (interjection *ha*, laughter) at seeing *Fur Breakfast* was colored by a series of kinesthetic attributive adjectives (*hairy* – 3 tokens – *whiskery*, *prickly to touch*). She reported being *quite disturbed* by the imaginary sensation of the fur on her skin (*touch and feel, the physicality of it*) and expanded on the negative emotions it provoked (*fear, disgust*).

(21) Emphi023: English speaker (reaction part)
Ha (laughs) this is like a hairy plate to the hairy bowl (laughs) a hairy brush (er) I don't like this very much because I don't like hairy things […]it's whiskery it looks like it would be prickly to touch oh my I wouldn't like to see it in real life this this one I find quite quite (er) disturbing probably more (laughs) than the others because of the physicality of it it's something you can actually touch and feel"[…] (er) a slight fear I think and a kind of disgust actually from the from the thought of it touching me.

Speakers also resort to the metonymic expression of pulling forces, loss of power and control, physical attraction and resistance. Destabilization is rendered by formulaic stock phrases such as *taken aback* (Emphi002, Emphi010), *caught off guard*

[29] Louise Bourgeois, *Spider* (1996).

(Emphi005), *this one is throwing me a bit* (Emphi008) and the subsequent loss of control is conveyed by such adjectives as *wobbly, dizzy, unstable* (Emphi012, Emphi023) or past participles like *drawn* and *sucked in* and allusions to an irresistible *force* (spirals or a tunnel in *Vision of Paradise*).

(22) Emphi002: English Speaker (reacting to *Vision of Paradise*):
I sort of feel sucked in cos of the spirals that are around the for
(23) Emphi004: Advanced Learner (reaction part)
Oh I'm drawn into it yes it's like a tunnel because (er) the manner you see it I felt drawn into it like there's a a force like a tunnel.

A French speaker (Emphi013) said she was amazed by this same painting ('émerveillée'), and that it reminded her of past dreams in which she fell and got lost ('des rêves où je tombais perdue'). Others emphasized their inability to react – utter stillness being one of the typical bodily responses of surprise – and used such words as *hypnotized*, ('captivé').

(24) Emphi001: Advanced learner (reaction part)
[...] *we are hypnotized and we have to to be to go *in the center of the painting but I don't want I want to to stay I want to stay out of it.*

These few examples – among many others – support the claim that metaphors and metonymies constitute conceptual representations and the basis of the lexical imagery relative to surprise and associated emotions, whether the speakers resort to standard existing formulae in their native language or struggle to come up with appropriate idioms in their second or foreign language.

4 Conclusion

In this article, we have shown that surprise presents invariant pivotal properties – disruption, disturbance, destabilization – followed by puzzlement and action-adjustment, which are not typical of other emotions, and are all evoked by the experiencers. This breach in common beliefs and expectations is first pre-conscious and then more explicitly conceptualized as speakers attempt to explain, rationalize and come to terms with the cognitive discrepancy. The mental processing is accessible through the analysis of a network of explicit and implicit language markers (both non-verbal and verbal ones) but requires the reconstruction of the subjacent cognitive scenario, as speech is intrinsically far less organized than written discourse. Surprise speech becomes even more disarticulated as the result of a sudden, unexpected disturbance, hence bouts of laughter, filled and blank pauses, repeats, unfinished sentences, shifts in pronouns and tenses. For the sake of categorization, we adopted the emotional and emotive concepts of the pragmatic theory of emotions, but all our examples show more complex configurations since expressive and descriptive discourse segments of surprise are not clearly distributed between spontaneous reaction phases and reporting phases but constantly merge into one another. The study also stressed the difficulty to isolate surprise from other emotions in speech (whether surprise is or is not an emotion is still pending), and we have repeatedly found examples in which two properties characteristic of emotions, *i.e.* valence and intensity, contributed to the subjective assessment of an episode as surprising. Thus we will state that associated emotions (which seem to be the norm rather than the exception) tend to have a cumulative influence upon the experience of surprise and that the more intense the emotion, the stronger the surprise. This may prove to be a more nuanced viewpoint in the ongoing debate whether surprise is neutral *per se* or positively and negatively valenced. Indeed, these partial results will have to be confirmed or qualified by analyzing the productions of a larger number of informants, which will lend itself to a more quantitative approach. Besides, it is worth examining other variables that can explain why the discourse of surprise – in a broad sense – can take so many forms. One has to do with the speaker's proficiency in the language (native *versus* non-native speakers) and the choice of a dominant language to express emotions for bilinguals. A subordinate question is whether native speakers are more likely to produce metaphoric and metonymic language, a main conduit for the language of surprise. Another interesting variable to explore is the connections between the properties of the source-event causing surprise and the personal history of the psychological subject (cf. Pavlenko 2006, 2008; Dewaele 2010, on the emotional perception of new events colored by the subject's autobiographical memory).

Appendix

Table 1 Main sources of surprise, series 1: 12 subjects

Green Man	**91.66%**	***n*10**
Slaughtered Pig	**75.00%**	***n*9**
Lost Paradise	**66.66%**	***n*8**
La Banquière	50.00%	*n*6
Disquieting Muses	41.66 %	*n*5
Pumpkin	33.33%	*n*4
Danae (Klimt)	16.66%	*n*2
Naked Man	16.66%	*n*2

Table 2 Main sources of surprise, series 2: 13 subjects

Implosion	**85.75%**	***n*12**
Janus Fleuri	**71.42%**	***n*10**
Carcass of Beef	**57.14%**	***n*8**
Illu	**57.14%**	***n*8**
Fur Breakfast	42.85%	*n*6
The Corn Field	7.14%	*n*1

References

Altarriba, Jeanette. 2006. Cognitive approaches to the study of emotion-laden and emotion words in monolingual and bilingual memory. In Pavlenko, A. (Ed.). *Bilingual minds, emotional experience, expression and representation.* Clevedon. England. Multilingual Matters: 232–256.

Bublatzky, Florian, Pedro Guerra, Carmen Pastor, Harald Schupp, and Jaime Vila. 2013. Additive effects of threat-of-shock and picture valence on startle reflex modulation. *PLOS ONE* 8 (1): 1–6 www.plosone.org, Last accessed Sept 2014.

Caffi, Claudia, and Richard Janney, eds. 1994. Involvement in language. *Special Issue of Journal of Pragmatics* 22: 325–373.

Cosnier, Jacques. 2006. *Psychologie des Émotions et des Sentiments.* Paris. France. Retz.

Damasio, Antonio. 1994. *Descartes' error: Emotion, reason, and the human brain.* New York: Grosset/Putnam Book.

Delancey, Scott. 2001. The mirative and evidentiality. *Journal of Pragmatics* 33: 369–382.

Dewaele, Jean-Marc. 2010. *Emotions in multiple languages.* London: Palgrave MacMillan.

Ekman, Paul. 1992. Are there basic emotions? *Psychology Review* 99: 150–153.

Goossens, Vannina. 2005. Les noms de sentiments, esquisse de typologie sémantique fondée sur les collocations verbales. *LIDIL* 32: 103–121.

Johnson-Laird, Peter, and Keith Oatley. 1989. The language of emotions: An analysis of a semantic field. *Cognition and Emotion* 3 (2): 81–123.

Kövecses, Zoltán. 2000. *Metaphor and emotion, language, culture and body in human feeling.* Cambridge: Cambridge University Press.

Lang, Peter, Margaret Bradley, and Bruce Cuthbert. 1990. Emotion, attention, and the startle reflex. *Psychological Review* 97 (3): 377–395.

Meyer, Wulf-Uwe, Rainer Reisenzein, and Achim Schützwohl. 1997. Toward a process analysis of emotions: The case of surprise. *Motivation and Emotion* 21 (3): 251–274.

Novakova, Iva, Vannina Goossens, and Elena Melnikova. 2012. « Associations sémantiques et lexicales spécifiques: sur l'exemple du lexique émotionnel des champs de surprise et de déception », *Congrès Mondial de Linguistique Française SHS Web of Conferences*, 1020–1021. http://www.shs-conferences.org/articles/shsconf/abs/2012/01/shsconf_cmlf12_000181/shsconf_cmlf12_000181.html. Last accessed on Sept 2014.

Ortony, Andrew, Gerald Clore, and Mark Foss. 1987. The referential structure of the affective lexicon. *Cognitive Science* 11: 341–364.

Ortony, Andrew, Gerald Clore, and Allan Collins. 1988. *The Cognitive Structure of Emotions.* Cambridge: Cambridge University Press.

Pavlenko, Aneta. 2006. Bilingual selves. In *Bilingual minds, emotional experience, expression and representation*, ed. A. Pavlenko, 1–33. Clevedon, Multilingual Matters.

———. 2008. Emotion and emotion-laden words in the bilingual lexicon. *Bilingualism: Language and Cognition* 11 (2): 147–164.

Plantin, Christian. 2011. *Les Bonnes raisons des émotions: Principes et méthode pour l'étude du discours émotionné.* Berne: Peter Lang.

Plutchik, Robert. 1980. *Emotion: A psychoevolutionary synthesis.* NewYork: Longman Higher Education.

Reisenzein, Rainer. 2000. The subjective experience of surprise. In *The message within: The role of subjective experience in social cognition and behavior*, ed. H. Bless and P. Forgas, 262–279. Philadelphia: Psychology Press.

Stein, Nancy, and Marc Hernandez. 2007. Assessing understanding and appraisals during emotional experience. In *Handbook of emotion elicitation and assessment*, ed. A. Coan and J. Allen, 298–317. Oxford: Oxford University Press.

Syssau, Arielle. 2006. Emotion et Cognition. In *Emotion et Cognition: Quand l'émotion parle à la cognition*, ed. N. Blanc et al., 13–67. Paris, Ed. In Press.
Valetopoulos, Freiderikos. 2013. Traduire la peur et la surprise: une étude contrastive en Grec moderne et en Français. In *Des Sentiments au point de vue*, ed. A. Chuquet, N. Raluca, and F. Valetopoulos, 95–116. Rennes: Presses Universitaires de Rennes.
Vermersch, Pierre. 1994. *L'Entretien d'explicitation en formation continue et initiale*. Paris: ESF.

Describing and Expressing Surprise

Agnès Celle, Anne Jugnet, Laure Lansari, and Emilie L'Hôte

Abstract The present paper aims to define the linguistic status of surprise, in comparison with other more prototypical emotions. Based on a sample of American spoken English (*In Treatment*), we argue that surprise entails an epistemic judgment whereby a specific event is connected to the speaker's set of expectations. Our contention is that surprise can be characterized as a disconnection between emotive and emotional communication (Marty A: Untersuchungen Zur Grundlegung Der Allgemeinen Grammatik Und Sprachphilosophie. Olms, Hildesheim/New York, 1976), identified linguistically as a disconnection between expression and communication (Bally C: Le Langage et La Vie. Payot, Paris, 1926). Emotional reactions of surprise are thus expressed by the speaker's intonation in mirative utterances, and by non-sentential elements in constructions. The second part of our study focuses on surprise lexemes (*surprise* (N) and *surprised* (Adj)). *Surprise* (N) cannot be classified as an epistemic state noun, nor as a prototypical psychological noun, as its prototypical sense is actually what we call its source reading. The state reading of the noun is thus identified as a metonymic extension of that prototypical sense, thereby reversing the expected structure of its semantic network. Similarly, the adjective *surprised* does not fall into prototypical scenarios, but rather rich and complex ones, pragmatically speaking. We show that the use of *surprised* (Adj) implies a comment on or an explanation of the initial emotional reaction, not the expression of it.

Keywords Surprise · Epistemic modality · Mirativity · Psychological nouns · Metonymy

Previous version of this chapter appeared in a work entitled "Expressing and describing surprise" – Celle, A. et al – pp. 215–244 in: "Expressing and Describing Surprise", Celle, A. and Lansari, L. eds. – (2017). Published by John Benjamins Publishing Company. Amsterdam/Philadelphia. https://benjamins.com/catalog/bct.92

A. Celle (✉) · A. Jugnet · L. Lansari · E. L'Hôte
Université Paris Diderot, Sorbonne Paris Cité, CLILLAC-ARP, EA 3967, Paris, France
e-mail: agnes.celle@univ-paris-diderot.fr

N. Depraz, A. J. Steinbock (eds.), *Surprise: An Emotion?*, Contributions
To Phenomenology 97, https://doi.org/10.1007/978-3-319-98657-9_11

1 Introduction

Surprise is one of the six primary emotions identified by Ekman (2003). Not all psychologists agree on whether surprise should be deemed an emotion, for surprise does not involve a valenced reaction.[1] However, there is a strong consensus regarding the temporality of the reaction of surprise (Depraz, Desmidt, this volume) as well as its manifestations. In linguistics, surprise is explored from different perspectives, not all of them being concerned with emotion as such. For instance, the verb *surprise* both in English and French is classified as a psychological verb by Levin (1993) and Mathieu (2000) on the basis of a syntactic feature, i.e. expressing the experiencer as the object. Recent accounts of the emotion lexicon have established correlations between the syntactic behavior of affect nouns and verbs and semantic dimensions such as intensity, aspect, causativity, and manifestation (Novakova et al. 2012). And yet, the goal of these studies is not to ascertain whether the surprise lexeme realized as a verb, a noun, or an adjective actually expresses emotion. Indeed, they all focus on the descriptive function of surprise lexemes but leave aside the expressive contribution of surprise words.

The aim of the present paper is to define the status of surprise with respect to emotion from a linguistic viewpoint. It is based on a sample of spoken dialogue data extracted from the American Series *In Treatment*, a TV drama about a psychologist conducting weekly sessions with patients. In this series, verbal interaction is part of a therapy in which patients discuss personal issues and interpersonal conflicts with their therapist in order to uncover problems and repair relationships. The patient-therapist relationship becomes more complex in the course of the therapy, as the therapist's self-doubt jeopardizes his position as a resolution expert.

On the basis of this data, and in accordance with Plantin (2011, 94; 114–115) and Michaelis (2001),[2] we uphold the view that emotion in general and surprise in particular cannot be reduced to a response to a stimulus. It is argued that surprise entails an epistemic judgment whereby a specific event is connected to the speaker's set of expectations. Drawing upon both the distinction between emotive communication and emotional communication originally made in philosophy by Marty (1976) and the parallel distinction between expression and communication originally made in linguistics by Bally (1926, 148), we argue that the most salient characteristic of surprise is to disconnect these two functions. Focusing on surprise lexemes, that is, on emotive rather than emotional communication, we then show that the noun *surprise* is not a prototypical psychological noun. Taking a closer look at the discourse use of the adjective *surprised*, we finally show that this adjective enters rich

[1] Ortony et al. (1988, 13) define emotions as "valenced reactions to events, agents or objects, with their particular nature being determined by the way in which the eliciting situation is construed."

[2] Michaelis (2001, 1039) argues that surprise entails a non-canonicity judgment and rightly points out that a physical startle may not be evidence of surprise: "For example, I might not have predicted a hallway encounter with a colleague, but I would not necessarily find that encounter surprising (even if my colleague startled me)."

pragmatic scenarios. Even if surprise lexemes do not express emotion, their use in discourse suggests that surprise is emotionally meaningful.

2 The Expression of Surprise

In an illuminating paper on how to develop a more unified pragmatics of emotive communication, Caffi and Janney (1994) stress that the foundations of such a project can be reconstructed from earlier contributions *inter alia* by Marty (1976) and Bally (1926) in the philosophy of language and in linguistics respectively. In an attempt to clarify the relation between affect and language, these scholars make a similar distinction between two communication modes. Marty (1976) distinguishes between emotional communication and emotive communication. Emotional communication is defined as "a type of spontaneous, unintentional leakage or bursting out of emotion in speech" (Caffi and Janney 1994, 328). Emotive communication, on the other hand, is "the intentional, strategic signaling of affective information in speech and writing [...] in order to influence partners' interpretations of situations and reach different goals" (Caffi and Janney 1994, 328). In the wake of Bergson,[3] Bally draws the line between expression and communication, the former being concerned with affect (*mode vécu*) and the latter with intellectual communication (*mode pur*). These categories, later taken up in philosophy of language and pragmatics, notably in Searle's (1969) theory of speech acts,[4] are highly relevant for philosophers, psychologists as well as linguists. Such categories provide linguists of different theoretical persuasions with a fruitful and powerful analytical tool for research on emotion.[5]

While emotions are certainly signaled by specific words in the lexicon, it is not clear whether emotion lexemes themselves actually express emotions, or whether they simply describe them. As pointed out by Legallois and François (2012: 209),

[3] Bergson's intellectual legacy in Bally's stylistics is pointed out by Legallois and François (2012, 198). These authors also highlight Bally's pioneering research on the relation between syntax and expressivity long before the role of emotion in discourse started being explored in cognitive linguistics as well as in construction grammar. Given the social nature of affective language in Bally's research, Caffi and Janney (1994, 335) even regard it as a first landmark in modern pragmatics.

[4] Searle (1969) distinguishes between assertive and expressive speech acts. In his theory, assertives aim to describe a state of affairs and may be judged true or false; expressives, on the other hand, express the sincerity condition of the speech act.

[5] Croft (2009, 412) argues that recognition of the speaker's intention by the addressee is a prerequisite for any successful linguistic communicative act, a concern that is not addressed in speech act theory: "the identification of construal as an essential part of symbolic meaning, and relating it to cognitive psychological processes, is a major contribution of cognitive semantics. But it must be situated in the larger model of language, communication and joint action in order to understand why it exists and how it is used by speaker and listener."

"expressivity is shown, it is not signified."[6] Typically, interjections seem to be devoted to the expression of emotions (*oh, hooray, wow*). But according to Kövecses (2000, 2–3), adjectives like *angry, happy, sad* and *depressed* "describe the emotion they signify" – while *I love you* may be used both to describe and express emotion. Like *angry, happy, sad* and *depressed*, the adjective *surprised* describes an emotion, but is not used to express it. Crucially, these two communication modes remain distinct in the case of surprise, which appears to be a characteristic of that emotion. While in Sect. 3, the function of surprise lexemes in the description of surprise is analyzed on the basis of our corpus data, we start this study with a brief outline of two major means of expressing surprise: mirativity (i) – a category that does not exist in the morphosyntax of English, and the use of specific constructions (ii).

2.1 Mirativity

Surprise is encoded in a number of languages through mirativity. Mirativity was first described as a category in its own right in DeLancey (2001, 369–370): "The term 'mirativity' refers to the linguistic marking of an utterance as conveying information which is new or unexpected to the speaker." DeLancey (1997, 44) illustrates the opposition between old first-hand information (1) and new information (2) in Lhasa Tibetan as follows:

(1) nga-r dngul to=tsam **yod**
I-LOC money some exist
'I have some money.' (e.g. I brought some with me)

(2) nga-r dngul to=tsam **'dug**
I-LOC money some exist
'I have some money!' (quite to my surprise)

The existential copula is the direct evidential *yod* in the case of old information (1), and mirative *'dug* in the case of unexpected new information (2).

It is not the purpose of the present paper to determine whether mirativity should be analyzed as a separate category or in relation to evidentiality. Let us note, however, that several scholars have argued for a unified treatment of information depending on speaker attitudes – rather than only on the source of information. By positing an epistemic scale, i.e. a continuum ranging from the realis domain to the irrealis domain, Akatsuka (1985) argues that surprise is related to the irrealis domain and suggests the following paraphrase for capturing the cognitive assimilation process of newly-learned information: "I didn't know this until this moment." More recently, Rett and Murray (2013, 457) have also accounted for the affinity between indirect evidence and speaker surprise by highlighting a common epistemic component. In

[6]"L'expressivité se montre, elle ne se signifie pas" (Legallois and François 2012, 209). Our translation.

Cheyenne, among other languages, the same morpheme may be interpreted either as a reportative, i.e. indirect evidential, or as a mirative evidential:

(3) a. É-x-hoo'k˙ohó-**neho**.
3-REM.PST-rain-NAR.SG.INAN
'It rained long ago, it is told.'
b. É-hoo'k˙ohó-**neho**!
3-rain-NAR.SG.INAN
'It's raining!' Rett and Murray (2013, 457)

The indirect evidential *neho* (3) is also a marker of mirativity (3), a fact that is widely represented across languages. Rett and Murray (2013, 470) argue that unlike direct evidentials in (1), indirect evidentials (3) and mirative markers in (2) and (3) both "relate the at-issue proposition p to some salient, contextually-valued set of expectations E."

Although mirativity does not exist as a morphosyntactic category in English, DeLancey (2001, 377–8) as well as Rett and Murray (2013) claim that it is marked by intonation. It is indeed possible in English to distinguish between a mirative intonational contour and the intonational contour of an ordinary statement.[7] A parallel may thus be drawn between the pair in (1) and (2) and that in (4), the difference between (a) and (b) being purely intonational:

(4) a. John arrived on time.
b. (Wow,) John arrived on time! (Rett and Murray 2013, 455)

According to Rett and Murray (2013, 467), following Rett (2011),[8] the mirative construction in (4) (b), conflates the assertion that p and the expression of speaker surprise: "Because the recency restriction is satisfied, the illocutionary content of the speech act is to assert that p and to express that p was not." Rett and Murray (2013, 455) put forward four distinguishing features of the expression of surprise. First, the expression of surprise is "undeniable in discourse" (# *No, you are not surprised*). Only the content can be denied (*No, John didn't arrive on time*). Second, it is "unembeddable under negation and other sentential operators." *John didn't arrive on time!* cannot express that the speaker is not surprised that John arrived on time; *Mary said John arrived on time!* can only express the speaker's surprise that Mary said John arrived on time, not the subject's surprise that John was on time. Third, the expression of surprise is always speaker-oriented – it has an exclamation intonation and not a question intonation. Fourth, it is subject to the recency restriction.

[7] DeLancey (2001, 377–378) and Rett and Murray (2013, 455–8) point out that the mirative intonational contour is often found in complimentary comments in English.

[8] Rett (2011, 413) defines sentence exclamations as both assertive and expressive: "The utterance of a sentence exclamation counts as an assertion of the denoted proposition p […] and an expression that p violates the speaker's expectation."

The first criterion, i.e. the deniability test, enables Rett (2011, 437) to highlight an important difference between the assertion of surprise and its expression, namely, that it is possible to confirm or deny an assertion, whereas the expression of surprise is not debatable:

(5) A: I am surprised that I won the contest.
B: Yes, you seem shocked.
(6) A: Wow, I won the contest.
B: # Yes, you seem shocked (Rett 2011, 437).

In Rett's words (2011, 436–437) "while the speaker's surprise in (5) can be agreed with, a sign that it was part of the assertion, the speaker's expression of unfulfilled expectation in (6) cannot."

Some of the criteria proposed by Rett to define the expressive dimension of mirative utterances are identical to those used by Potts to define the term *expressive* (2005, 156). Potts draws a parallel between expressives and conventional implicatures. According to him, it is not clear whether they are best conceived as entailments. Potts argues that expressives have the properties of lexicality, entailment, speaker orientation, and independence.[9] On this view, the deniability test reveals that it is difficult to determine whether expressives are usual entailments. "Expressives are, in this sense, performative. But expressives, unlike more familiar entailments, are also not challengeable by a hearer." For instance, "a sincere utterance of *damn* cannot be challenged or turn out to be false" (Potts 2005, 157). Similarly in (7), the property *bastard* attributed by Speaker A (Ann) cannot be challenged by Speaker B (Kyle) in the following example:

(7) Ann: That bastard Conner was promoted.
Kyle: Conner is not a bastard (Potts 2005, 157).

Although the pairing of the subject and the predicative complement is denied by negative *not*, Potts claims that "the contribution of the epithet *that bastard* does become part of the common ground. Kyle refuses to accept the characterization, but this does not in any way mitigate Anne's use of the epithet." In Potts's view, expressives express properties of the speaker's emotional states – hence outright denial of their content by a hearer makes little sense.

[9] Potts gives the following definitions for these properties. Lexicality: "expressive meaning is part of the lexical meaning of certain expressions, a semantic quality of words and phrases (Löbner 2002, 32). Entailment: "the aspects of meaning under discussion, in particular, the semantic information displayed by expressive, can have consequences for the notion of logical validity" (Kaplan 1999, 13). Speaker orientation: "another characteristic distinguishing expressive meaning from propositional meaning is that it is valid only for the utterer, at the time and place of utterance. This limitation it shares with, for instance, a smile, a frown, a gesture of impatience […]" (Cruse 1986, 272). Independence: "Expressive meaning carried by a lexical item in a statement plays no role in determining its truth-conditions." (Cruse 1986, 272)

The distinction made by Rett (2011) and Rett and Murray (2013) following Potts (2005) is an essential one. Not only do these authors provide formal arguments to disentangle assertion and expression of surprise, but their results also allow us to better understand why the adjectival phrase "be surprised" may enter pragmatic scenarios (see Sect. 4 p. 17). Assertion as such (*I am surprised*) may be denied or debated in dialogue, allowing for surprise to be (self-)attributed. In contrast, the expression of surprise is indexical and not accessible to the hearer, who may only deny the content: in (6), *No, you didn't win the contest* would be a felicitous reply by B, as opposed to #*No, you are not surprised.* However, we maintain that Rett's argument should be qualified. Even in (5), B's reply is necessarily a modalised assertion (*yes, you* ***seem*** *shocked*) because it is not possible for B to produce an assertion about A's inner state (*#yes, you are surprised*). The infelicity of (6) should be reconsidered, given that the following would be more felicitous than (6) as recorded above:

(6) A: Wow, I won the contest.
B: You sound shocked/surprised.

Continuation with an evidential utterance in which speaker B evaluates whether s/he rightly perceived A's expressive statement is perfectly possible, as long as the utterance is not an outright confirmation or a blunt denial of a proposition. This pattern is well-represented in our corpus data, as in (8):

(8) Frances: I kept telling myself, "don't worry, Frances, it's like riding a bicycle."
Paul: And was it?
Frances: He took off my shirt and touched me, and I thought I was gonna lose it. I mean, **he loved my breasts**.
Paul: **You sound surprised**.
Frances: I'm not 25 or synthetic, if you know what I mean. And he just kept kissing them and touching them. Well, I forgot how sensitive they are. Are you shocked? (*In Treatment*)

On the basis of its intonational contour in the film and despite the absence of an exclamation mark in the script, *he loved my breasts* is taken to be a sentence exclamation, i.e. a mirative utterance. While it is true that this utterance cannot be agreed with, which is a sign that it differs from a straightforward assertion, the evidential utterance that follows functions as a request for A's confirmation of B's construal of A's expressive statement. Thus although the proposition <he – loved my breasts> is not stable enough to be confirmed or denied, the identification of the emotion that is associated with it is mediated by the auditory evidential.

In Sects. 3 and 4, we focus on the noun *surprise* and on the adjectival phrase *be surprised*, i.e. on lexemes which describe, rather than express surprise. However, it may turn out to be difficult to draw a dividing line between expressives and assertives. For instance, surprise lexemes may be found in exclamative constructions, which, according to Rett (2011, 2008), are expressive, as in (9):

(9) What a surprise! How surprising!

Our corpus data suggest that in such cases, surprise lexemes are generally used ironically. Along Rett's (2011, 416; 2008, 603) line of reasoning, such uses correspond to insincere speech acts. We contend, however, that ironic expressives are not only a matter of insincerity. From a psychological point of view, ironic expressives illustrate that expressives do not necessarily coincide with emotional communication. Contra Potts (2005), we argue that expressives do not necessarily reflect the speaker's emotional state. Following Legallois and François (2012, 210–211), we consider expressivity to be able to serve not only an emotional function, but also a mimetic one – as in onomatopoeia – and an ethical one – typically in self-presentation and enactment constructions such as *be + like*. In the latter case, expressives contribute vividness to argumentation, allowing the speaker to display an emotion that s/he does not feel. Our hypothesis is that the ethical function of expressives is exploited in emotive communication. From a linguistic point of view, we show in Sect. 3 that this use might be better accounted for in terms of modal/evaluative judgment than insincerity. In addition, we argue that the noun *surprise* used in verbless utterances and exclamations does not behave like a prototypical psych noun. Predicative and metonymic uses of the noun *surprise* lead us to offer a reinterpretation of its semantic network altogether.

2.2 Constructions

Constructions are the second major means of expressing surprise. To our knowledge, two types of construction have been described in the literature. Firstly, Mad Magazine sentences (Akmajian 1984; Lambrecht 1990) and Incredulity Response Constructions (Fillmore et al. 1988: 511) are types of exclamatives that border on interrogative sentences: "MMs are used by speakers to express surprise, disbelief, skepticism, scorn and so on, at some situation or event" (Akmajian 1984: 2). Secondly, the "What's X doing Y?" (WXDY) construction is defined by Kay and Fillmore (1999: 21) in this way: "In the WXDY construction we have an overall scene (frame) in which a judgment of incongruity is attributed by a judge (prag) to some state of affairs." Such constructions are not represented in the corpus data chosen for this study. Therefore, we will simply mention their contribution to the expression of surprise in connection with mirativity. The WXDY construction is illustrated in the following example:

(10) Waiter, **what's this fly doing in my soup?** (Kay and Fillmore 1999: 4)

Such a question is not a request for information about the activity of the fly, but a request for an explanation of its presence in the soup. As shown by Kay and Fillmore (1999: 30), the interrogative *what* is not referential here. Kay and Fillmore call this use of WXDY a construction because like a "how come" question, it "directly encodes, in addition to a request or demand for an explanation, the pragmatic force of attributing what we call INCONGRUITY to the scene or proposition for which the explanation is required" (Kay and Fillmore 1999: 4). "How come" questions are represented in our corpus data. The proposition that is questioned may not be repeated when it is mentioned in the preceding context:

(11) Patient: I'm not ready to talk about that yet.
Paul : **How come?** (*In Treatment*, III 18)

As a question, this construction differs from mirative utterances in that it expresses surprise without asserting it openly. Furthermore, this expression of surprise is not indexical, but semantically inherent to the construction. Unlike mirative utterances, WXDY constructions allow for the sense of incongruity to be embedded, as in a conversation between members of a criminal defense team in (X):

(12) How are we going to deal with our client's confusing account of **what the photographs were doing in his brief case**? (Kay and Fillmore 1999: 5)

As shown by Kay and Fillmore (1999: 5; 21-22), the source of the incongruity judgment is unclear in embedded contexts: "This means that the WXDY construction does not tell the interpreter how to identify the source of the judgment of incongruity but, in effect, contains an instruction that this is something one should be able to figure out for oneself." The judgment is "conventionally associated" with the construction, which accounts for its unrestricted use with respect to time and source. By contrast, a mirative utterance is subject to the recency restriction (cf. Rett) and can only be interpreted as such in an indexical relation to the speaker:

(13) Waiter, there's a fly in my soup!

The other construction type that expresses surprise takes the form of an incredulous quotation of a context proposition[10] in interrogative or exclamative utterances:

(14) Speaker A: I hear that John may wear a tuxedo to the ball…
Speaker B: **Him wear a tuxedo?** He doesn't even own a clean shirt (Lambrecht 1990: 216).

[10] Lambrecht (1990, 220) distinguishes between the context proposition, i.e. the proposition that precedes in the immediate context, and the context sentence in which the argument and the predicate of the context proposition are quoted and called into question. We use the same terminology for the sake of clarity. However, the term "context sentence" is not well-chosen since it designates a non sentential utterance.

This use of the infinitive was first described by Akmajian (1984) and compared with the imperative. Lambrecht (1990) claims that this type of structure made up of a NP and a predicate is related not to the imperative, but to topic-comment constructions. He argues that the quotation can only be analyzed correctly from a constructionist perspective, by looking at the preceding context proposition and at the follow-up proposition in which the speaker's judgment is expressed. The context sentence forms a complex topic, whereas the follow-up proposition is a comment about that topic, more specifically a rejection of its validity. Lambrecht (1990: 222) further argues that the IRC is part of a widely attested construction which he calls the Unlinked-Topic Construction, from which it differs in only one respect: "in the IRC one of the topic constituents (the NP) designates a discourse entity."

The context sentence is indeed an incredulous reaction to a discourse event, deconstructing as it does the very pairing of the argument and the predicate previously made by Speaker A. This construction expresses surprise, but unlike mirative utterances it challenges the validity of p. In addition, emotion is expressed in stages: Firstly, the context sentence abstracts away the predication from any temporal situation and subjective source, sounding purely emotional. It expresses surprise and incredulity, which is reflected in its non-sentential syntax.[11] Secondly, a syntactically more elaborate comment introduces Speaker B's point of view. It is only in that comment that Speaker B's emotive judgment appears, overtly rejecting the validity of the context proposition.

Constructions express surprise without asserting that p. The expression of surprise may either be inherent to the construction (WXDY construction) or arise from the pragmatic use of a topic-comment construction. In English, it appears that the emotional reaction of surprise is expressed by intonation in mirative utterances and by non sentential elements in constructions. Surprise lexemes are combined neither with mirative utterances nor with constructions. This leads us to focusing more specifically on the noun *surprise* and its uses in context.

3 The Noun *Surprise*: An Atypical Psych Noun

The noun *surprise* has often been characterized as a 'psych' noun, i.e. a noun that can describe the emotional state of an experiencer. Our aim is to determine whether this noun is indeed best characterized as a lexeme describing an emotion, i.e. whether this noun shares the properties of core members of the class (such as *love*, *hate* or *disappointment*).

In various languages, psych nouns can be polysemous, in that they can either denote a state or the source (or stimulus) of the state. *Disappointment*, *embarrass-*

[11] Our analysis of the emotional contribution of this construction differs from Lambrecht's. Lambrecht regards the follow-up judgment as emotional. We argue that the expression of the speaker's point of view in the follow-up proposition is emotive rather than emotional. By contrast, no point of view is associated with the context sentence, which conveys a spontaneous emotional reaction.

ment, *preoccupation* or *obsession* (*inter alia*) can denote a feeling or the state of affairs that triggers this feeling (Barque et al. 2012; Tutin 2009). (15) illustrates the feeling of reading *disappointment* (*that feeling of disappointment* presupposes *disappointment is what you are feeling*), while (16) illustrates the source reading (*a disappointment* is understood as *a source of disappointment*):

(15) That feeling of disappointment comes from you (*In Treatment*, III.14).
(16) Was I that big a disappointment to you? (*In Treatment*, I.5)

Our aim is to determine whether *surprise* illustrates this regular polysemy, i.e. whether it belongs to the subclass of polysemous (psych) nouns that can both denote a state and have 'derived' meanings and refer to the objects or sources of an affect. Our analysis is based on the attested uses of the noun *surprise* in the script of the series *In Treatment*. When evidence was lacking in our *In Treatment* corpus (henceforth IT), we present examples taken from a preliminary study done on the occurrences of the noun in the spoken section of the Corpus of Contemporary American English (henceforth COCAs). The occurrences of *surprise* (N) were encoded on the basis of Plantin's definition of emotional utterances. Plantin (2011) argues that an emotional utterance combines a series of four parameters: an emotion term, an experiencer, a situation and an attributor. We thus analyzed the occurrences of the noun *surprise*, specifying whether the noun had a state or a source (/object) interpretation (emotion term), whether the experiencer was mentioned or identifiable in context (experiencer), whether the source or stimulus was mentioned (situation), and which attributor could be identified.

We discovered that though the opposition between stimulus and state corresponds to an ontologically clear distinction, it is often difficult to make a clear distinction between the state reading and the source reading of the noun in context. It has been argued that psych nouns tend to be mass nouns when they denote a state while they tend to be used as count nouns when they denote a stimulus. Tutin (2009, 68), however, qualifies this generalization, arguing that independent parameters – such as the modification by a qualifying adjective or a type reading, may lead to the use of psych nouns as count nouns. This mass/count criterion can be combined with a second generalization: When the noun describes the source of a state, it cannot occur with a complement describing (specifying) this source:

(17) The early arrival of their son [= source]
was a surprise to them [=experiencer].
(18) A couple welcomed the surprisingly early arrival of their new son en-route to hospital vs.*A couple welcomed the surprise <u>of the early arrival of their new son</u> [noun surprise denoting the source + of-complement specifying it].

But as soon as we look at language in use – as is the case in this study for instance, it is often much less obvious whether a state or a source reading is to be preferred, in spite of these two theoretical criteria.

3.1 The State/Source Polysemy – Ambiguities

The difficulty to disambiguate the state/source readings is partly due to the fact that the noun *surprise* is very often used predicatively, or appears in set phrases. This use, as is argued below, is usually correlated with a comment or epistemic judgment on a given state of affairs (rather than with the expression of an emotion). Though the prototypical function of noun phrases is to refer to objects and to be the arguments of a predicate, some nouns can also be used predicatively, to describe or to comment on a referential argument. Predicative uses of the noun *surprise* are prominent in our corpus data: In the eleven occurrences of *surprise* in IT, the noun does not refer to a given participant in a situation, but is used to characterize a situation that was either previously mentioned or mentioned in the following clause.

Various types of predicative uses can be distinguished. Most often the noun is part of the predicate, and is introduced by the copula *be*, as in (19), (20) and (21):

(19) Of course they are that's not really a big surprise (IT II,1).
(20) But they didn't help, which is not really a surprise because the health center is total shit (IT II,2).
(21) It's no surprise that I can't find a man (IT II,31).

The noun also appears in verbless sentences – without any determiner, modifier or complement. In these cases we consider that the noun is used predicatively, as we assume that verbless sentences correspond to a syntactic structure which involves an verbless predicate – whether adjectival, nominal, pronominal, adverbial or prepositional – associated with (assertive, exclamative, interrogative or imperative) modality (cf. Lefeuvre (1999) *inter alia*). The modal character of verbless sentences is reflected in the presence of parameters usually found in exclamative utterances, such as repetitions as in (22), or the 'interjection'/morpheme *fucking* in (23).

(22) *Surprise, surprise* (IT I,4).
(23) *Surprise, sur-fucking-prise* (IT II,27).

The noun also occurs in the set phrase *by surprise*, as in (24), which we classify as an instance of predicative use. In this case *surprise* does not refer to a participant's state of mind in a given situation, but is part of a modifying prepositional phrase, easily paraphrased by an adverbial phrase such as *unawares*.

(24) So your feelings kind of caught you by surprise (IT II, 6).

The predicative uses of a noun make it difficult to determine whether a state or a source reading is to be preferred, as the noun is not used to 'refer' to a stative eventuality or an object, but rather to comment on a topic. We have nonetheless chosen to determine whether the noun is closer to a state or a source reading, as predicates (whether verbal, adjectival or nominal) can also be polysemous. We thus assume that the distinction between state and source readings still holds in predicative contexts. We now examine in more detail the three types of occurrences of the noun *surprise* in IT, focusing on the state/source ambiguity.

3.2 Frequent Source Readings

The analysis of specific occurrences of *surprise* leads us to conclude that its source reading is relatively more frequent than the state reading.[12]

When the noun is part of the verb phrase, in extraposed constructions, we give it a source reading: as illustrated in (25), (26) and (27), *surprise* can be replaced with *a cause for surprise* in these three contexts.

(25) [cf. (19)] Of course they are that's not really a cause for a big surprise/a reason to be much surprised.

(26) [cf. (20)] But they didn't help, which is not really a cause for surprise because the health center is total shit.

(27) [cf. (21)] It's no cause for surprise that I can't find a man.

In these cases, *surprise* is used as a count noun, as is often the case when it has a source meaning. It is noteworthy that negation receives a particular interpretation in such contexts: as the noun is used to comment on a given event, what is negated is not the occurrence of the event but rather its characterization as a cause for surprise. This event can be characterized as not being a (prototypical) member of the class of surprising events (according to the speaker's mental representations), as in *it is not a surprise...*, or it can be characterized as lacking all features of a surprising event (according to the speaker's standards), as in *It is no surprise that I can't find a man.*

In verbless utterances such as (22) above, it is particularly difficult to favour one interpretation over the other. Our first criterion, i.e. the use of the noun as a mass noun, may justify its interpretation as a state noun. But it is difficult to perform our second test (using a more explicit noun phrase, such as "*#Feeling of surprise, feeling of surprise*" or "*#Cause for surprise, cause for surprise*"). It could be argued that verbless sentences such as "surprise, surprise" are used to describe the addressee's expectations, i.e. that a state reading is to be preferred. But this type of verbless sentences can also be used ironically (as in (28)), in which case a possible paraphrase for such utterances would be: "(It is) no cause for surprise."

(28) I have a theory about that pizza. - Surprise, surprise. - Would you like to hear it? - Do I have a choice? (IT, I,28)

---

[12] This conclusion corroborates the findings of a previous study of 200 occurrences of the noun *surprise* in the COCAs.

In another context, i.e. the prepositional phrase *by surprise* in (24), the noun *surprise* can be argued to have a state reading. Our first criterion is that it is used as a mass noun (the singular noun is introduced by the zero determiner rather than the indefinite determiner *a*). Second, though it is difficult to use our second criterion, i.e. the replacement with a more explicit noun phrase (*#your feelings caught you by a state of surprise/#your feelings caught you by a cause for surprise*), the possible paraphrases of this set phrase clearly involve the description of a state, as in (29):

(29) Your feelings caught you unawares/in a state of surprise.

Though the interpretation of the noun in the set phrase *by surprise* is not straightforward, we therefore argue for a state interpretation. What is not obvious is whether the state of the subject described by such a phrase is an emotional state or rather an epistemic state.

We consider that the difficulties to disambiguate the reading of *surprise* in many contexts is significant, as they point to this noun differing from usual psych nouns (such as *sorrow* or *disappointment*), as well as from other nouns related to the notion of surprise. If the noun *wonder* seems to be as ambiguous as its near-synonym *surprise*, nouns like *amazement*, *astonishment*, *bewilderment* or *stupefaction* are on the other hand non-ambiguous, simple state nouns, denoting the (episodic) state of a subject. The fact that they are not ambiguous may be related to their denoting an 'extreme' state, or a high degree of surprise. In any case, the ambiguity of *surprise* is identified as a crucial property of the noun: At the very essence of *surprise* (N) lies the fact that the state and source readings cannot be analysed as two clearly distinct facets in its semantics, because the two interpretations are too closely connected. Such a position is also justified by the fact that some features of one reading can be connected to the other interpretation. For instance in (30), the noun *surprise* gets a source interpretation - as shown by the use of the indefinite determiner *a* and the possible paraphrase *That's not really a cause for surprise*.

(30) Well, of course the Navy are going to absolve himself. - Of course they are that's not really a big surprise (IT II,1).

However, the modifier *big* cannot modify the source, as it cannot modify events or specific pieces of information. It is rather used to describe the intensity of the state of surprise: as shown in Morzycki (2009), size adjectives can get a degree interpretation when they modify gradable nouns. So *surprise* (N), even when receiving a source reading, can be associated with degree modifiers that usually modify state nouns.

Having established that the two facets of meaning are intrinsically related, we now examine more precisely the type of relation that holds between them, which allows us to revise the semantic network of the noun.

3.3 *The State Reading as a Metonymic Extension*

Previous descriptions of polysemous state nouns rely on the often implicit assumption that the source reading is derived from the state noun (Novakova et al. 2012; Tutin 2009). More specifically, it is often hypothesized that the source reading is a metonymic extension of the state noun (Barque et al. 2012). Our analysis of attested uses of *surprise* (N) from a cognitive perspective leads us to question the validity of such a hypothesis. Based on the definition of conceptual metonymy and its attested properties in language in use, we argue that *surprise* (N) cannot be considered as a prototypical state noun.

Cognitive linguistics considers metonymy as a conceptual phenomenon with a linguistic expression: "Metonymy is a conceptual projection whereby one experiential domain (the target) is partially understood in terms of another experiential domain (the source) included in the same common experiential domain" (Barcelona 2003, 4). Adopting a cognitive approach to the semantic network of *surprise* (N) involves the identification of a prototype and of a series of senses that are extensions of this prototype: "A prototype is a typical instance of a category, and other elements are assimilated to the category on the basis of their perceived resemblance to the prototype; there are degrees of membership based on degrees of similarity" (Langacker 1987, 371). In his seminal work on the issue, Langacker only allows for metaphorical extensions, but more recent publications have come to emphasize the relevance of metonymic extensions in semantic networks (Seto 2003; Blank 2003; Radden and Kövecses 1999).

Cognitive semantics provides us with several arguments in favor of a semantic network for the noun *surprise* in which the source reading is considered as the prototypical reading, while the state reading becomes a metonymic extension of the latter. These arguments rely on etymology (i), on cognitive patterns involved in metonymic relations (ii), on the distinction between predicative and referential use (iii), and on the relevance of set expressions and linguistic patterns in the analysis of metonymic extensions (iv). In this study, these arguments are only used to determine how the source reading and the state reading of surprise are connected in the semantic network of the noun – not to prove whether or not there is a metonymic extension within the network in the first place, which is one of the prerequisites of our work.

Let us start with the etymological argument (i). Diachronically speaking, the source reading of *surprise* is older than its state reading in English. In the late fourteenth century, occurrences of the noun are found to mean "unexpected attack or capture." Similarly, in 1590, surprise occurs with the meaning of "something unexpected." It is only in 1600 that attested use of the state reading can be found (Hoad 1993). While recent work in metaphor analysis (Pragglejaz Group 2007) points out that the basic (and thus non-metaphorical) reading of a word is not necessarily the oldest, we argue that in the case of the analysis of metonymic relations within a semantic network, the diachronic argument is relevant.

In their work on the development of a comprehensive and integrated cognitivist view of metonymy (ii), Radden et Kövecses (1999, 39) show that consequences of an event are more frequently identified as metonymies of its causes than the other way around. This corroborates our hypothesis that in the case of *surprise*, it is the state reading – the state of surprise that is triggered by a surprising event – that is a metonymic extension of the *source reading* – the surprising event which causes a subsequent state of surprise.

The distinction between referential and predicative uses of a noun is the third argument we put forward in favor of our hypothesis (iii). Indeed, in his study of the role of mappings and domains in the understanding of metonymy, Ibanez and José (2003, 114) points out that a metonymy is rarely used predicatively, but tends to be used referentially. These findings fit within our analyses, which have established that a majority of predicative uses of the noun *surprise* rely on the source reading, while the state reading rarely occurs in such cases. Thus it makes sense to consider the source reading as prototypical, and the state reading as a metonymic extension of the latter.

Finally, in his corpus-based study of the occurrences of the noun *eye*, Hilpert (2006, 145) argues that all the metonymic extensions identified for the semantic network of the word occur as part of recurrent linguistic patterns (iv). Similarly, in the case of *surprise*, we have noted that set expressions and recurring linguistic patterns (e.g. *by surprise*) account for a large proportion of the occurrences of the state reading of the noun, thereby pointing to the latter following the usual pattern of metonymic extensions rather than prototypical readings for nouns.

Based on the combination between our preliminary COCAs study, the analysis of our IT corpus data and the aforementioned series of arguments, we thus propose a gradient of meaning between the prototypical sense of surprise (i.e. its source reading) and its metonymic extension (i.e. its state reading). This allows us to take into account the attested ambiguities of the noun in the new representation of its semantic network, primarily because of inherent ambiguities in the definition of domains in cognitive linguistics: according to Barcelona (2003, 8), they include "all the entrenched knowledge that a speaker has about an area of experience." This means that just like prototypes, domains are culturally-dependent and may even vary from speaker to speaker. Just as the distinction between metaphor and metonymy can sometimes be blurred by these fuzzy boundaries between domains, we argue that the distinction between the prototypical sense of a word and its metonymic extension is not always clear-cut for similar reasons. This is how we account for the fact that *surprise* (N) collocates with adjectives usually associated with states as well as with adjectives usually associated with events. All previously analyzed occurrences of *surprise* are thus placed on a gradient of meaning between the source reading of the noun (i.e. its prototypical sense) and its state reading (i.e. the metonymic extension of the prototype). Cognitive semantics proposes a specific location on the source-state gradient, thereby allowing for ambiguities to be fully accounted for.

Several arguments have led us to claim that the prototypical sense (or 'primary reading') of the noun *surprise* is its source reading, its state reading being a metonymical extension. In this respect, *surprise* differs from most other psych nouns like *sorrow*, *worry*, *concern*, *obsession* or *disappointment*, which primarily denote states and not sources. This conclusion leads us to wonder whether surprise can be analyzed as a true psych noun.

3.4 What Kind of State?

We thus question the very inclusion of *surprise* in the class of psych nouns in English, as it differs from 'prototypical' (polysemous) state nouns, which tend to follow the patterns described by Barque et al. (2012): they can denote states or sources, but are not particularly ambiguous, as in (31).

(31) She gets overwhelmed. She needs to vent. So after she vents and she puts the stress, the <u>disappointment</u> and the <u>worry</u> of her own life on your shoulders, who do you turn to? (IT II,17)

The noun is usually not used to describe the feelings or emotions of an experiencer, but rather an epistemic or cognitive judgment on the likelihood of a given state of affairs. As illustrated above (examples (19) to (21)), surprise is often found in the extraposed structure (i.e. the pattern *it is (not) a surprise that p*). The function of such a structure is not to describe the feelings of an experiencer, but rather to introduce a judgment as to the (un)expected character of a situation. It could thus be argued that the structure allows a speaker to express an epistemic judgment (based on a cognitive state, a given situation is considered (un)likely), rather than an emotional stance towards a state of affairs. The structure *it's no surprise that p* can thus be paraphrased as *According to my knowledge (set of expectations), that p is very likely/obvious/must be the case*. The interpretation of verbless sentences (examples (22) and (23)) is similar to that of the extraposed structure: The speaker judges a given situation likely (or unlikely), and does not express his/her emotion. Even in contexts in which the noun can have a state reading, as *by surprise* in (24), the noun is not used to express the cognitive or epistemic state of a subject (as shown by the paraphrase *unawares*: such an adverb characterizes a subject's stance on a topic, not their own cognitive state).

Still, *surprise* shares some properties with typical psych nouns. Like other state nouns, *surprise* can appear in the prepositional phrase *in + his/her* (or any description of an experiencer) + psych noun. In this phrase, the noun denotes a feeling with a causal value - or, in Van de Velde's (1998) terms, the feeling is conceived as an (abstract) location of a subject.

(32) a. In his fear, he reached out for support./In his despair he cursed those who had taken her away from him.
b. In his surprise, he could not help staring, suddenly sure that the fellow Klingon would recognize and expose him.

Surprise can also be the argument of *experience* or *feel:*

(33) The despair/fear/surprise that he experienced…

But contrary to Blumenthal's analysis (2009) for instance, we argue that the lexeme *surprise* is not a central (or prototypical) psych noun, based on a series of parameters.

We also reject the hypothesis that *surprise* is better characterized as an epistemic state noun. If this hypothesis were validated, *surprise* (and *wonder*) would be similar to nouns such as *belief, knowledge* or *assumption*, which can also be ambiguous and denote the state or the (propositional) content of an epistemic stance. Such nouns, which are a subclass of "conceptual shells" (Schmid 2000), can have sentential complements, and can appear in an identificational structure involving a sentential constituent (the clause describing a propositional content), as in (34) and (35):

(34) Still, the idea that we have little access to the workings of our minds is a dangerous one.
(35) The idea that she mentioned is that we have little access to the workings of our minds is a dangerous one.

There is no attested use of the noun *surprise* in such contexts in IT or in COCAs. Moreover, if *surprise* denoted an epistemic stance or its content, it could be modified by the adjectives *true* or *false* (cf. *a false belief/assumption/hypothesis*), which is obviously not the case. Thus *surprise* cannot be considered a prototypical epistemic state noun either.

Overall, *surprise* cannot be considered as a prototypical psych noun, but cannot be classified as an epistemic state noun either. In other words, the noun resists precise categorization. The specificity of surprise lexemes is confirmed by our analysis of the adjective *surprised.* Section 4 investigates the discourse use of the adjective *surprised* in association with first and second persons in the present tense.

4 The Adjective Surprised and Complex Pragmatic Scenarios

This corpus study confirms that surprise lexemes do not express emotion. The adjective *surprised* enters various pragmatic scenarios which all point to a disconnection between the assertion and the expression of surprise. Two main cases can be distinguished in the patient/therapist interactions under scrutiny in our IT corpus. First, the adjective appears in questions asked or inferential statements made by patients (*You're surprised, Are you surprised?*) in an attempt to expose the therapist's surprise – more often than not the therapist responds by denying his surprise in order to regain control over the situation. Secondly, the adjective is used in assertions of surprise (*I'm surprised*). Such assertions do not correspond to an expression of surprise but constitute a strategy to minimize emotions such as anger or disappointment, or correspond to a way of rationalizing surprise.

4.1 Experiencing and Controlling One's Surprise

We first focus on interactions where the therapist is clearly surprised by one of his patients' declarations and yet refuses to acknowledge his surprise, thus illustrating what Plantin (2011, 150) calls "dissonance" between emotive and emotional communication.[13] Such instances show that attributing surprise may be viewed as face-threatening and, as a corollary, that displaying surprise may amount to a problematic loss of control. As described in Goffman's work on the issue, hiding or denying one's surprise is thus a face-saving attempt to regain control (Goffman 1986).

In his work on the figurative language of emotions, Kövecses (2000) argues that the metaphors used to express emotions all rely on the same cognitive scenario involving five stages:

Cause ➔ Emotion ➔ Control ➔ Loss of Control ➔ Behavioral Response

Kövesces does not assume that this scenario governs the other two types of emotion language he defines (i.e. descriptive language and non-metaphorical expressive language), but the control stage is particularly salient in our IT corpus. The need for control on the part of the therapist is highly understandable in the context of his sessions with his patients. Surprise is inherently epistemic and emerges when there is a discrepancy between one's expectations and reality, thus exposing the speaker's "unprepared mind" (Aikhenvald 2006, 195). For the therapist, being caught off-guard by his patients would point to his possible lack of insight on the matter discussed, as exemplified in (36):

(36) Laura: Because he's right. Andrew's right. I am being unfaithful to him.
Paul: You mean what happened last night at the club?
Laura: Not just last night. I've been unfaithful to him for a long time.
Paul: <u>Why haven't we talked about this before?</u>
Laura: I think we have. It's been here all along. You mean to say you've never noticed it? **You're surprised.**
Paul: **No, I'm not surprised.** <u>I just...I don't follow what you mean.</u>
Laura: This isn't... the reaction I was hoping for. Not at all (IT).

Laura has been in love with her therapist, Paul, for a long time and has decided to tell him. Seeing Paul's face, Laura realizes he is surprised by her tentative revelation. Her inferential statement "you're surprised" may be regarded as an indirect reproach, hence Paul's denial (*No, I'm not surprised*). His denial is all the more

[13] "dissonance entre l'émotif et l'émotionnel." Our translation.

interesting as it is contradicted by his facial reaction throughout the dialogue and by linguistic evidence, namely the syntactic disfluency *I just… I don't follow….*

Similarly, in (37), Paul didn't expect Alex to come, and Alex infers from Paul's face that he is surprised. However, Paul is reluctant to admit his surprise:

(37) Alex*:* ***Surprised?***
Paul*: I didn't think, uh... I didn't think you'd come back again.*
Alex*: I'm back.*
Paul*: Good. Come on in. Would you like some coffee? I've just made a pot.*
Alex*: Yeah, coffee's good* (IT).

In both examples, Paul denies or eludes his surprise by resorting to rational arguments – he points to his lack of understanding in (36) and refers to prior expectations in (37) – but his surprise remains visible at the syntactic level. In (38), as Paul almost admits to being surprised, syntactic disfluency in his response is associated with the clear acknowledgment that surprise is not an adequate reaction on his part (see *I shouldn't be*):

(38) Alex: OK, so my squadron commander calls me back up to his office. This time all the brass were there and they want to know if I'm interested in active duty. If I am, I have to go through a modified basic training at Pensacola, a battery of tests, physical, psychological... just to make sure I've recovered. So I said, "Bring it on, baby! I'm ready for anything!" **Are you surprised?**
Paul: Well, a little, I guess. I, um... I shouldn't be, but I suppose it was always a possibility.
Alex: But you think it's too fast, right?
Paul: Do you think it's too fast?
Alex: Doc, you have no idea how good it feels to be useful again. I think that's been the real problem. I've been sitting around feeling sorry for myself. I'm a doer, not a talker, so all of this, that's over. Those days are over.

In all these utterances, the speaker experiencing surprise never spontaneously asserts his surprise by saying *I am surprised.* This points to the linguistic expression of surprise being more syntactic than lexical, as evidenced by the recurrence of questions and cases of disfluency. Conversely, we argue that saying *I'm surprised* does not correspond to the expression of an emotional reaction of surprise either.

4.2 *Asserting One's Surprise*

4.2.1 Minimizing One's Anger and Disappointment

Our contention is that, in many cases, saying *I'm surprised* does not constitute an expression of surprise but rather a strategy to minimize other emotions associated with negative valence such as anger and disappointment. Resorting to a more neutral emotion[14] is thus a way to save one's face and/or avoid hurting the interlocutor's feelings. Like Paul's reluctance to admit surprise examined in previous examples, this strategy of minimization is an attempt to control one's emotions. Contrary to (36) and (37) above, no hesitation may be observed in the utterance supposedly expressing surprise in the following dialogue between Paul and Gina, her supervisor:

(39) Gina: You've created this cold version of me that's limiting and castrating. Because when you're arguing with me you want to see yourself as battling the forces of repression. And if the truth is that I'm an emotional person who has to rein herself in, it doesn't fit with what you need, now does it? [...] I didn't sleep with Charlie because... that's not who I wanted to be. I loved David and I loved the life we had and who I was with him. I loved myself with David. David's Gina. And you know what? When I was standing over...his grave at the funeral...I thought... if I was magically reborn... I would choose him all over again. For all his faults...(Sniffs)...he allowed me to be who I wanted to be. **I'm surprised you didn't see it, how crazy I was about him**. He was everything.
Paul: I never heard you say it.
Gina: Oh, shit. (Sniffs) Everything has to be said? And now I have to sit here and listen to your crappy, near-sighted, self-involved theories about my life.

Gina's "*I'm surprised you didn't see it, how crazy I was about him*" is not spontaneously triggered by what Paul has just said; rather, it comes as a well-prepared conclusion to her own angry speech. Gina relies on the surprise lexicon to reproach Paul with his misconception of who she is. It is an indirect way to tell him: *you should have seen it*. The syntactic structure in the complementation of *surprised* (right dislocation) explicitly points to the referent of the pronoun *it* (i.e. *how crazy I was about him*), thereby suggesting a lack of trust in Paul's own analysis. *Surprised* does not denote emotional communication but emotive communication here. What Gina is truly experiencing is more akin to anger and disappointment – and more particularly person-related disappointment.[15] The question is, then, why she should say *I'm surprised* instead of stating *I'm disappointed/angry/sad...*, which are defined as negatively valenced adjectives in the literature on emotions (Ortony et al. 1988). Let us compare (39) with (40), in which there is no denial, since pregnant Amy admits to experiencing negative emotions:

(40) Paul: Have you spoken to Dr Kornreich about terminating this pregnancy that you've wanted for so long?

[14] See Ortony *et al.* (1999, 32), who point out that "surprise can arise in the absence of a valenced reaction."

[15] See Van Dijk and Zeelenberg (2002) on the distinction between outcome-related disappointment and person-related disappointment. These authors contend that anger is always person-related.

Amy: Yes. I don't think, um... I'll be able to love another baby. **I'm angry at him already**. I feel like it's forcing itself into the middle of things. Ever since I became pregnant I felt the hormones raging, driving me insane, and I know I'll have postpartum, that's what happened last time.
Paul: **Are you angry? It feels like you're angry tonight.**
Amy: I'd say that's a fairly accurate assessment.
Paul: You wanna tell me why?
Amy: Actually, **I'm angry at myself** for taking your advice, even though I knew better. I will say that it's not the first time **I've left here disappointed or angry.**
Paul: If I was you, I'd look into that.
Amy: Of course, **I'm angry at everything these days**. More or less all the time. This week was particularly...

The therapist's suggestion (*If I was you, I'd look into that*) shows that negative emotions are inherently problematic and are proof of a malfunction. Asserting person-related disappointment or anger in (39) would have been a direct reproach on Gina's part, and the sign of her problematic relationship with Paul. In terms of modality, I'm angry/disappointed that you didn't see it denotes root modality,[16] whereas *I'm surprised* presents itself as an epistemic judgment and amounts to *I don't understand*: it merely states the discrepancy between one's prior expectations and reality, without any negative evaluation. In other words, *I'm surprised* is part of Gina's control strategy to sound as professional as possible and to avoid an open conflict with Paul. But after his justification (*I never heard you say it*), she finally gives way to her anger, thus illustrating the shift from control to loss of control defined in Kövesces's scenario.

The same kind of control strategy is at stake in the exchange between Paul and his patient Mia in (41). Mia, who has always been a difficult patient, accuses Paul of sleeping with another patient. Paul tries to remain calm and refuses to acknowledge that he is angry with her for interfering with his private life:

(41) Mia: Paul, do you have any idea how many therapists are successfully sued for sleeping with their patients?
Paul: What are you trying to say, Mia?
Mia: How could you? Was she that young and pretty? Or did she make you feel that big and important? Take it. It's yours. Do you want me to pick that up? It's all yours. I've just made an ass of myself, haven't I? I bet you're angry.
Paul: Maybe I am. Mostly, **I'm... I'm surprised.**
Mia: No, you're furious. I can see it all over your face. But... you just can't express it because you're not supposed to. You want me to go?

[16] Contrary to epistemic modality, root modality is concerned with the relationship between the speaker and another subject. It involves modal uses such as obligation, permission, ability.

The hesitation in Paul's verbal response shows that he is buying time to find an appropriate label for his reaction. The choice of surprise, a neutral emotion which is not associated with a negative valence, is nonetheless contradicted by his facial reaction. Mia is no fool and is fully aware of Paul's control and face-saving strategy. She points out that admitting his anger would be unprofessional on his part (cf. *because you're not supposed to*). It would clearly reveal that Mia has managed to destabilize him – which is exactly what she has been aiming at since the beginning of her therapy.

It would be tempting to conclude from the examples examined so far that the assertion of surprise never corresponds to an expression of surprise but is systematically used to hide or deny negative emotions. Our data suggest however that the "dissonance" between emotive and emotional communication is more complex than that.

4.2.2 Rationalising Surprise

In (42), it is not the therapist who says *I'm surprised* but a patient – Amy, who had a miscarriage during a previous session with Paul.

(42) Amy: Where's the stain? I saw it when we left here. You must've scrubbed the couch like a lunatic.
Paul: Actually, it, uh... it came off quite easily.
Amy: Oh.
Paul: Hm.
Amy: **I'm surprised,** I... actually thought you'd have been down on all fours scrubbing away, and your wife standing over you shouting instructions and... you saying, "It's not coming out, I don't know what to do, "my next patient'll be here any minute." I didn't think it would come out that easily, that's all.
Paul: Did you want it to still be there?
Amy: Yes, I was very attached to it. Where's your sense of humour? C'mon.
Paul: Well... Guess I, uh... I don't find it so... so funny. I think you're disappointed.
Amy: Why are you making a big deal out of it?
Paul: What were you expecting to feel when you saw it?
Amy: Nothing.
Paul: No sadness? No grief?
Amy: Over a stain?
Paul: No. Over the loss of your baby. Maybe you wanted to see that stain because you needed to accept it, and all the mixed feelings you have about it.

The interjection *oh* indicates that Amy is surprised by the fact that the blood stain on Paul's couch has disappeared. Contrary to examples (39) and (40) examined above, there seems to exist here a correspondence between the lexeme used by the speaker and the emotion experienced by the latter. It is worth noticing, however, that Amy's first verbal response to the surprising element is not an assertion of surprise but a spontaneous, uncontrolled emotional outburst (*oh*). This points to the fact that *I'm surprised* may be uttered to express surprise but only once surprise has been cognitively integrated – as though *I'm surprised* corresponds to a comment on or an explanation of the initial truly emotional reaction. So although no control strategy is involved in the case of (42), it is still questionable whether *I'm surprised* should be analysed as expressing surprise. We contend that it is best analysed as an attempt to rationalize surprise, as it is indeed immediately followed by expressions pointing to the discrepancy between the subject's expectations and reality (*I actually thought/I didn't think*). It is also relevant that Paul questions Amy's surprise and reinterprets it as disappointment (*I think you're disappointed*). He thus suggests that Amy is denying her true feelings to protect herself. As a therapist, he cannot believe that her miscarriage has not brought about negatively valenced emotions.

5 Conclusion

The goal of this paper was to establish what the linguistic status of surprise was, in comparison with other more prototypical emotions, none of which are considered to be reducible to the response to a stimulus (Plantin 2011; Michaelis 2001). Based on our analyses of dialogues taken from the US TV series *In Treatment*, we have argued that surprise entails an epistemic judgment whereby a specific event is connected to the speaker's set of expectations. According to our study, surprise can be characterized as the locus of a disconnection between emotive and emotional communication (Marty 1976), identified linguistically as a disconnection between expression and communication (Bally 1926). This means that linguistic constructions express surprise without stating it explicitly. The expression of surprise may either be inherent to a given construction, or stem from the pragmatic use of a topic-comment construction. In our data, emotional reactions of surprise are expressed by the speaker's intonation in mirative utterances, and by non-sentential elements in constructions. As we have found that surprise lexemes are combined neither with mirative utterances nor with constructions, the second part of our study has concentrated on the former, i.e. *surprise* (N) and *surprised* (Adj) in our *In Treatment* data. S*urprise* (N) is shown to resist clear-cut categories. It cannot be classified as an epistemic state noun, nor as a prototypical psychological noun, as its prototypical sense is actually what we call its source reading. The state reading of the noun is thus identified as a metonymic extension of that prototypical sense, thereby reversing the expected structure of its semantic network, based on those associated with most other psych nouns like *sorrow, worry, concern, obsession* or *disappointment*. Similarly, the adjective *surprised* does not fall into prototypical scenarios, but rather rich and complex ones, pragmatically speaking. A speaker usually does not rely on the phrase *I'm surprised* to express her/his surprise, but uses it as part of a strategy aimed at playing down negatively valenced emotions like anger or disappointment. The reliance on a more neutral emotion like that of surprise thus becomes a face-saving strategy and/or an attempt to control one's emotions. We point out that *I'm surprised* can only be used to express surprise once that emotion has been cognitively integrated by the speaker. The reliance on *surprised* (Adj) then becomes a comment on or an explanation of the initial emotional reaction, not the expression of it.

Thus our study goes beyond the idea that the assertion of surprise is systematically used to hide or deny negative emotions. The dissonance identified between emotive and emotional communication is presented as more complex in our data. Similarly, even if surprise lexemes do not express emotion, their use in discourse still suggests that surprise is emotionally meaningful.

Acknowledgement This study was funded by the Agence Nationale de la Recherche as part of the Emphiline Project – EMCO (Emotion(s), Cognition, Comportement) programme under contract number ANR-11–EMCO-0005.

References

Aikhenvald, Alexandra Y. 2006. *Evidentiality*. Oxford: Oxford University Press.

Akatsuka, Noriko. 1985. Conditionals and the epistemic scale. *Language* 61 (3): 625–639.

Akmajian, Adrian. 1984. Sentence types and the form-function fit. *Natural Language and Linguistic Theory* 2: 1–23.

Bally, Charles. 1926. *Le Langage et La Vie*. Paris: Payot.

Barcelona, Antonio, ed. 2003. *Metaphor and metonymy at the crossroads: A cognitive perspective*. Topics in English Linguistics 30. Berlin: Mouton de Gruyter.

Barque, Lucie, Antonio Fábregas, and Rafael Marín. 2012. *Les Noms D'état Psychologique et Leurs" Objets": Étude D'une Alternance Sémantique*. Lexique. https://129.242.170.252/munin/bitstream/handle/10037/4603/article.pdf?sequence=3.

Blank, Andreas. 2003. Polysemy in the Lexicon and in Discourse. In *Polysemy: Flexible patterns of meaning in mind and language*, ed. Brigitte Nerlich, Zazie Todd, Vimala Herman, and David Clarke D., 142:267–96. Trends in linguistics. Studies and monographs, vol. 142. Berlin/New York: Mouton de Gruyter.

Blumenthal, Peter. 2009. Les Noms D'émotion: Trois Systèmes D'ordre. In *Le Lexique Des Émotions*, ed. Iva Novakova and Agnès Tutin, 41–64. Langues, Gestes, Paroles. Grenoble: ELLUG, Université Stendhal.

Caffi, Claudia, and Richard W. Janney. 1994. Toward a pragmatics of emotive communication. *Journal of Pragmatics* 22 (3-4): 325–373. https://doi.org/10.1016/0378-2166(94)90115-5.

Croft, William. 2009. Toward a social cognitive linguistics. In *New directions in cognitive linguistics*, ed. Vyvyan Evans and Stephanie Pourcel, 24:395–420. Human Cognitive Processing. Amsterdam/Philadelphia: John Benjamins Pub. Co.

Cruse, D. A. 1986. *Lexical semantics*. Cambridge textbooks in linguistics. Cambridge [Cambridgeshire]. New York: Cambridge University Press.

DeLancey, Scott. 1997. Mirativity: The grammatical marking of unexpected information. *Linguistic Typology* 1 (1).

———. 2001. The mirative and evidentiality. *Journal of Pragmatics* 33 (3): 369–382.

Depraz, Natalie and Thomas Desmidt. This volume.
Fillmore, Charles J., Paul Kay, and Mary Catherine O'Connor. 1988. Regularity and idiomaticity in grammatical constructions: The case of 'Let Alone. *Language* 64 (3): 501–538.
Goffman, Erving. 1986. *Frame analysis: An essay on the organization of experience*. Boston: Northeastern University Press.
Hilpert, Martin. 2006. Keeping an eye on the data: Metonymies and their patterns. In *Corpus-based approaches to metaphor and metonymy*, ed. by Anatol Stefanowitsch and Stefan Thomas Gries, 123–51. Trends in Linguistics. Studies and Monographs, vol. 171. Berlin: M. de Gruyter.
Hoad, T. F. 1993. *The concise Oxford dictionary of English etymology*. Oxford Paperback Reference. Oxford/New York: Oxford University Press.
Ibanez, Ruiz de Mendoza, and Francisco José. 2003. The role of mappings and domains in understanding metonymy. In *Metaphor and metonymy at the crossroads: A cognitive perspective*, Topics in English Linguistics 30, ed. Antonio Barcelona, 109–132. Berlin: Mouton de Gruyter.
Kaplan, David. 1999. *What is meaning? Explorations in the theory of meaning as use*. Ms. Draft 1. UCLA.
Kay, Paul, and Charles J. Fillmore. 1999. Grammatical constructions and linguistic generalizations: The what's X doing Y? construction. *Language* 75 (1): 1–33.
Kövecses, Zoltán. 2000. *Metaphor and emotion: Language, culture, and body in human feeling*. Studies in Emotion and Social Interaction. Second Series. Cambridge/New York/Paris: Cambridge University Press; Editions de la Maison des sciences de l'homme.
Langacker, Ronald W. 1987. *Foundations of cognitive grammar: Theoretical prerequisites*. Stanford: Stanford University Press.
Lambrecht, Knud. 1990. "What, me worry?" – 'Mad Magazine Sentences' revisited. Annual meeting of the Berkeley linguistics. *Society* 16 (1): 215–228.
Lefeuvre, Florence. 1999. Les 'Marqueurs de Prédication' Dans La Phrase Averbale En Français. *Verbum* XXI, 4 (XXI, 4): 429–438.
Legallois, Dominique, and Jacques François. 2012. Définition et Illustration de La Notion D'expressivité En Linguistique. In *Relations, Connexions, Dépendances: Hommage Au Professeur Claude Guimier*, ed. Claude Guimier, Nicole Le Querler, Franck Neveu, and Franck Roussel, 197–221. Rennes: Presses universitaires de Rennes.
Levin, Beth. 1993. *English verb classes and alternations: A preliminary investigation*. Chicago: University of Chicago Press.
Löbner, Sebastian. 2002. *Understanding semantics*. Understanding language series. London: Arnold.
Marty, Anton. 1976. *Untersuchungen Zur Grundlegung Der Allgemeinen Grammatik Und Sprachphilosophie*. Hildesheim/New York: Olms.
Mathieu, Yvette Yannick. 2000. *Les Verbes de Sentiment: De L'analyse Linguistique Au Traitement Automatique. Sciences Du Langage*. Paris: CNRS éditions.
Michaelis, Laura. 2001. Exclamative constructions. In *Language typology and language universals: An international handbook*, ed. Martin, 1963- Haspelmath, 1038–50. Handbücher Zur Sprach- Und Kommunikationswissenschaft. Berlin/New York: W. de Gruyter.
Morzycki, Marcin. 2009. Degree modification of gradable nouns: Size adjectives and adnominal degree morphemes. *Natural Language Semantics* 17 (2): 175–203.
Novakova, Iva, Vannina Goossens, and Elena Melnikova. 2012. Associations sémantiques et syntaxiques spécifiques. Sur l'exemple du lexique émotionnel des champs de surprise et de déception. In *3e Congrès mondial de linguistique française: Lyon, 4-7 juillet 2012*, ed. Franck Neveu, Valelia Muni Toke, and Peter Blumenthal, 1017–29. Les Ulis: EDP Sciences.
Ortony, Andrew, Gerald L. Clore, and Allan Collins. 1988. *The cognitive structure of emotions*. Cambridge/New York/New York: Cambridge University Press.
Plantin, Christian. 2011. *Les Bonnes Raisons Des Émotions: Principes et Méthode Pour L'étude Du Discours Émotionné*, vol. 94, Sciences Pour La Communication. Bern/New York: Peter Lang.
Potts, Christopher. 2005. *The logic of conventional implicatures*. Oxford: Oxford University Press.

Pragglejaz Group. 2007. MIP: a method for identifying metaphorically used words in discourse. *Metaphor and Symbol* 22 (1) 1.

Radden, Günter, and Zoltán Kövecses. 1999. Towards a theory of metonymy. In *Metonymy in language and thought*, ed. Klaus-Uwe, 1942- Panther and Günter Radden, v. 4:17–59. Human Cognitive Processing, 1387–6724; v. 4. Amsterdam/Philadelphia: J. Benjamins

Rett, Jessica. 2008. A degree account of exclamatives. In *Proceedings of SALT* 18:601–18. http://elanguage.net/journals/salt/article/download/18.601/1921.

———. 2011. Exclamatives, degrees and speech acts. *Linguistics and Philosophy* 34 (5): 411–442.

Rett, Jessica, and Sarah Murray. 2013. A semantic account of mirative evidentials. In *Proceedings of SALT*, 23:453–472. http://elanguage.net/journals/salt/article/view/23.453/0.

Schmid, Hans-Jörg. 2000. *English abstract nouns as conceptual shells. From corpus to cognition.* Berlin/New York: Mouton de Gruyter.

Searle, John. 1969. *Speech acts: An essay in the philosophy of language*. London: Cambridge University Press.

Seto, Ken-Ichi. 2003. Metonymic polysemy and its place in meaning extension. In *Polysemy: flexible patterns of meaning in mind and language*, ed. Brigitte Nerlich, Zazie Todd, Vimala Herman, and David Clarke D., 142:195-216. Trends in linguistics. Studies and monographs; 142. Berlin/New York: Mouton de Gruyter.

Tutin, Agnès. 2009. Les Émotions Sont-Elles Comptables? In *Le Lexique Des Émotions*, ed. by Iva Novakova and Agnès Tutin, 55–79. Langues, Gestes, Paroles. Grenoble: ELLUG, Université Stendhal.

Van de Velde, Danielle. 1998. Cet Obscur Objet Du Désir. L'objet Des Verbes de Sentiment. *Travaux de Linguistique* 35: 67–78.

Van Dijk, Wilco W., and Marcel Zeelenberg. 2002. What do we talk about when we talk about disappointment? Distinguishing outcome-related disappointment from person-related disappointment. *Cognition & Emotion* 16 (6): 787–807.

The manufacturer's authorised representative in the EU is Springer
Nature Customer Service Centre GmbH, Europaplatz 3, 69115 Heidelberg,
Germany. If you have any concerns regarding our products, please
contact ProductSafety@springernature.com

Printed and bound by CPI Group (UK) Ltd, Croydon, CR0 4YY
05/10/2026
02228131-0001